Use R!

Series Editors:
Robert Gentleman   Kurt Hornik   Giovanni Parmigia

Use R!

Pfaff: Analysis of Integrated and Cointegrated Time Series with R

Bernhard Pfaff

Analysis of Integrated and Cointegrated Time Series with R

With 19 Figures

Springer

Bernhard Pfaff
Ludwig-Sauer-Strasse 7
D-61476 Kronberg im Taunus
GERMANY
bernhard@pfaffikus.de

Series Editors:
Robert Gentleman
Program in Computational Biology
Division of Public Health Sciences
Fred Hutchinson Cancer Research Center
1100 Fairview Ave. N, M2-B876
Seattle, Washington, 981029-1024
USA

Kurt Hornik
Department für Statistik und Mathematik
Wirtschaftsuniversität Wien Augasse 2-6
A-1090 Wien
Austria

Giovanni Parmigiani
The Sidney Kimmel Comprehensive Cancer Center at Johns Hopkins University
550 North Broadway
Baltimore, MD, 21205-2011
USA

Library of Congress Control Number: 2005935027

ISBN-10: 0-387-27959-8   e-ISBN 0-387-27960-1
ISBN-13: 978-0387-27959-6

Printed on acid-free paper.

Printed in the United States of America.   (SBA)

9 8 7 6 5 4 3 2 1

springeronline.com

To Toni

Preface

This book's title is the synthesis of two influential and outstanding entities. To quote David Hendry in the Nobel Memorial Prize lecture for Clive W. J. Granger, "[the] modeling of non-stationary macroeconomic time series [...] has now become the dominant paradigm in empirical macroeconomic research" [43]. Hence, a thorough command of integration and cointegration analysis is a must for the applied econometrician. On the other side is the open-source statistical programming environment R. Since the mid-1990s it has grown steadily out of infancy and can now be considered a mature, flexible and powerful software with more than 600 contributed packages. However, it is fair to say that R has by now not received the attention among econometrician it deserves. This book tries to bridge this gap by showing how easily the methods and tools encountered in integration and cointegration analysis are implemented in R.

This book adresses senior undergraduate and graduate students and practioneers alike. Although the book's content is not a pure theoretical exposition of integration and cointegration analysis, it is particularly suited as an accompanying text in applied computer laboratory classes. Where possible, the data sets of the original articles have been used in the examples such that the reader can work through these step by step and thereby replicate the results. Exercises are included after each chapter. These exercises are written with the aim of fostering the reader's command of R and applying the previously presented tests and methods. It is assumed that the reader has already gained some experience with R by working through the relevant chapters in Dalgaard [13] and Venables and Ripley [98] as well as the manual "An Introduction to R."

This book is divided into three parts. In the first part, theoretical concepts of time series analysis, unit root processes, as well as cointegration are presented. Although the book's aim is not a thorough theoretical exposition of these methods, this first part serves as a unifying introduction to the notation used and as a brief refresher of the theoretical underlyings of the practical examples in the later chapters. The focus of the second part is the testing of the unit root hypothesis. The common testing procedure augmented Dickey–Fuller test for detecting the order of integration is considered first. In the later sections other unit root tests encountered widely in applied econometrics, such as the Phillips–Perron, Elliott–Rothenberg–Stock, Kwiatkowski–Phillips–Schmidt–Shin and Schmidt–Phillips tests, are presented as well as the case of seasonal unit roots and processes that are contaminated by structural shifts. The topic of the third and last part is cointegration. As an introduction, the two-step method by Engle and Granger and the method proposed by Phillips and Ouliaris are discussed before finally the Johansen's method is exemplified. The book ends with an exposition of vector-error-correction models that are affected by a one-time structural shift.

At this point I would like to express my gratitude to the R Core Team for making this software available to the public and to the numerous package authors who have enriched this software environment. The anonymous referees are owed a special thanks for the suggestions made. Of course, all remaining errors are mine. Last but not least, I would to thank my editor, John Kimmel, for his continous encouragement and support.

Kronberg im Taunus,   *Bernhard Pfaff*
                                                        September 2005

Contents

Part III Cointegration

Part I

Theoretical Concepts

1

Stationary Autoregressive-Moving Average (ARMA) Processes

Although this book has integration and cointegration analysis as its theme, it is nevertheless a necessity to first introduce some concepts of stochastic processes as well as the stationary ARMA model class. Having paved this route, the next steps, i.e., the introduction of nonstationary, unit root-, as well as long memory-processes, will follow.

1.1 Characteristics of Time Series

A discrete *time series*[1] is defined as an ordered sequence of random numbers with respect to time. More formally, such a *stochastic process* can be written as

$$\{y(s,t), s \in \mathfrak{S}, t \in \mathfrak{T}\}, \tag{1.1}$$

where for each $t \in \mathfrak{T}$, $y(\cdot, t)$ is a random variable on the sample space $\mathfrak{S}$ and a realization of this stochastic process is given by $y(s, \cdot)$ for each $s \in \mathfrak{S}$ with regard to a point in time $t \in \mathfrak{T}$. Hence, what we observe in reality are realizations of an unknown stochastic process, the *data-generated process*:

$$\{y\}_{t=1}^{T} = \{y_1, y_2, \ldots, y_t, \ldots, y_{T-1}, y_T\}, \tag{1.2}$$

with $t = 1, \ldots, T \in \mathfrak{T}$.

One aim of time series analysis is concerned with the detection of this data-generated process by inferring from its realization to the underlying structure. In the Figure 1.1, the path of real U.S. Gross National Product (GNP) is depicted.[2] By mere eyespotting a "trend" in the series is evident. By comparing the behavior of this series with the unemployment rate for the same time span, *i.e.*, from 1909 until 1988, a lack of a "trend" is visible.

This artifact leads us to the first characteristic of a time series, namely, *stationarity*. The ameliorated form of a stationary process is termed *weakly stationary* and is defined as

---

[1] The first occurence of subject entries are set in *italics*.

[2] The time series are taken from the extended Nelson and Plosser [69] data set (see Schotman and van Dijk [89]).

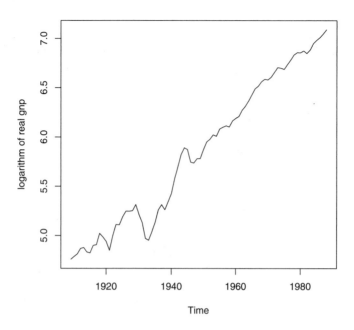

Fig. 1.1. U.S. GNP (real, logarithm)

$$E[y_t] = \mu < \infty, \forall t \in \mathfrak{T} \, , \qquad (1.3a)$$

$$E[(y_t - \mu)(y_{t-j} - \mu)] = \gamma_j, \forall t, j \in \mathfrak{T} \, . \qquad (1.3b)$$

Because only the first two theoretical moments of the stochastic process have to be defined and being constant, finite over time, this process is also referred to as being *second-order stationary* or *covariance stationary*. In that sense, the real U.S. GNP would not qualify as a realization of a stationary process because of its trending nature. Whether this is also the case for the U.S. unemployment rate, has to be seen.

Aside from weak stationarity, the concept of a *strictly stationary* process is defined as

$$F\{y_1, y_2, \ldots, y_t, \ldots, y_T\} = F\{y_{1+j}, y_{2+j}, \ldots, y_{t+j}, \ldots, y_{T+j}\} \, , \qquad (1.4)$$

where $F\{\cdot\}$ is the joint distribution function and $\forall t, j \in \mathfrak{T}$. Hence if a process is strictly stationary with finite second moments, then it must be covariance stationary as well. Although stochastic processes can be set up to be covariance stationary, it must not be a strictly stationary process. It would be the case, for example, if the mean and autocovariances would not be functions of

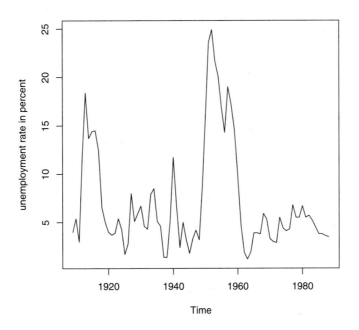

Fig. 1.2. U.S. unemployment rate (in percent)

time but of higher moments instead.

The next characteristic of a stochastic process to be introduced is *ergodicity*.[3] Ergodicity refers to one type of asymptotic independence. In prose, asymptotic independence means that two realizations of a time series become ever closer to independence, the further they are apart with respect to time. More formally, asymptotic independence can be defined as

$$|F(y_1, \ldots, y_T, y_{j+1}, \ldots, y_{j+T}) - F(y_1, \ldots, y_T)F(y_{j+1}, \ldots, y_{j+T})| \to 0 \,, \quad (1.5)$$

with $j \to \infty$. The joint distribution of two subsequences of a stochastic process $\{y_t\}$ is equal to the product of the marginal distribution functions the more distant the two subsequences are from each other. A stationary stochastic process is ergodic if

$$\lim_{T \to \infty} \left\{ \frac{1}{T} \sum_{j=1}^{T} E[y_t - \mu][y_{t+j} - \mu] \right\} = 0 \,, \quad (1.6)$$

---

[3] For a more detailed discussion and definition of ergodicity, the reader is referred to Davidson and MacKinnon [14], Spanos [94], White [100], and Hamilton [38].

holds. This equation would be satisfied if the autocovariances tend to zero with increasing j.

Finally, a *white noise* process is defined as

$$E(\varepsilon_t) = 0 \,, \tag{1.7a}$$

$$E(\varepsilon_t^2) = \sigma^2 \,, \tag{1.7b}$$

$$E(\varepsilon_t \varepsilon_\tau) = 0 \quad \text{for} \quad t \neq \tau \,. \tag{1.7c}$$

When necessary, ε_t is assumed to be normally distributed: $\varepsilon_t \backsim \mathcal{N}(0, \sigma^2)$. If Equations 1.7a–1.7c are amended by this assumption, then the process is said to be a *normal-* or *Gaussian white noise* process. Furthermore, sometimes Equation 1.7c is replaced with the stronger assumption of independence. If this is the case, then the process is said to be an *independent white noise* process. Please note that for normally distributed random variables, uncorrelatedness and independence are equivalent. Otherwise, independency is sufficient for uncorrelatedness but not *vice versa*.

1.2 AR(p) Time Series Process

We start by considering a simple *first-order autoregressive process*. The current period's value of $\{y_t\}$ is explained by its previous one, a constant c, and an error process $\{\varepsilon_t\}$.

$$y_t = c + \phi y_{t-1} + \varepsilon_t \,, \tag{1.8}$$

where $\{\varepsilon_t\}$ obeys Equations 1.7a–1.7c, *i.e.*, is a white noise process. Basically, Equation 1.8 is a first order inhomogenous difference equation. The path of this process depends on the value of ϕ. If $|\phi| \geq 1$, then shocks accumulate over time and hence the process is nonstationary. Incidentally, if $|\phi| > 1$, the process grows without bounds, and if $|\phi| = 1$ is true, the process has a *unit root*. The latter will be discussed in more detail in Section 2.2. In this subsection, however, we will only consider the covariance stationary case, *i.e.*, $|\phi| < 1$. With the lag operator L, Equation 1.8 can be rewritten as

$$(1 - \phi L)y_t = c + \varepsilon_t \,. \tag{1.9}$$

The stable solution to this process is given by an infinite sum of past errors with decaying weights:

$$y_t = (c + \varepsilon_t) + \phi(c + \varepsilon_{t-1}) + \phi^2(c + \varepsilon_{t-2}) + \phi^3(c + \varepsilon_{t-3}) + \ldots \tag{1.10a}$$

$$= [\frac{c}{1 - \phi}] + \varepsilon_t + \phi\varepsilon_{t-1} + \phi^2\varepsilon_{t-2} + \phi^3\varepsilon_{t-3} + \ldots \,. \tag{1.10b}$$

It is left to the reader as an exercise to show that the expected value and the second-order moments of the AR(1)–process in Equation 1.8 are given by

$$\mu = E[y_t] = \frac{c}{1 - \phi} \, , \tag{1.11a}$$

$$\gamma_0 = E[(y_t - \mu)^2] = \frac{\sigma^2}{1 - \phi^2} \, , \tag{1.11b}$$

$$\gamma_j = E[(y_t - \mu)(y_{t-j} - \mu)] = [\frac{\phi^j}{1 - \phi^2}]\sigma^2 \tag{1.11c}$$

(see Exercise 1). By comparing Equations 1.3a–1.3b with 1.11a–1.11c, it is clear that the AR(1)–process $\{y_t\}$ is a stationary process. Furthermore, from Equation 1.11c, the geometrically decaying pattern of the autocovariances is evident.

In the Rcode example 1.1, a stable AR(1)–process with 100 observations and $\phi = 0.9$ is generated as well as a time series plot, its *autocorrelations* and *partial autocorrelations* as bar plots.[4] In Figure 1.3, the smooth behavior of

---

Rcode 1.1 Simulation of an AR(1)–process with $\phi = 0.9$

```
set.seed(123456)                                                        1
y <- arima.sim(n = 100, list(ar = 0.9), innov=rnorm(100))               2
op <- par(no.readonly=TRUE)                                             3
layout(matrix(c(1, 1, 2, 3), 2, 2, byrow=TRUE))                         4
plot.ts(y, ylab='')                                                     5
acf(y, main='Autocorrelations', ylab='', ylim=c(-1, 1))                6
pacf(y, main='Partial Autocorrelations', ylab='', ylim=c              7
   (-1, 1))
par(op)                                                                 8
```

$\{y_t\}$ caused by a value of ϕ close to one is visible. Also, the slowly decaying pattern of the autocorrelations is clearly given. The single spike at lag one in the partial autocorrelations indicates an AR(1)–process.

The AR(1)–process can be generalized to an AR(p)–process:

$$y_t = c + \phi_1 y_{t-1} + \phi_2 y_{t-2} + \ldots + \phi_p y_{t-p} + \varepsilon_t \, . \tag{1.12}$$

---

[4] In this Rcode example, functions contained in the standard package **stats** are used. However, it should be pointed out that the same functionalities are provided in the contributed CRAN package **fSeries** by Würtz et al. [102]. These functions include the simulation (**armaSim()**), estimation (**armaFit()**) and prediction (**predict()**) of autoregressive integrated moving average (ARIMA)–models as well as stability evaluation (**armaRoots()**) and the calculation of theoretical autocorrelation and partial autocorrelation functions (**armaTrueacf()**). Furthermore, S3-methods for summaries, printing, and plotting accompany these functions. The advantage for the user using these functions is given by a coherent argument list across all functions.

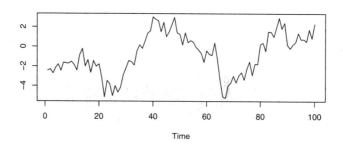

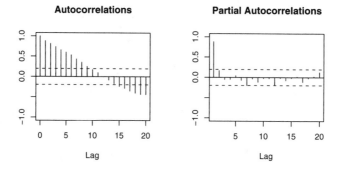

Fig. 1.3. Time series plot of AR(1)–process, $\phi = 0.9$

Likewise to Equation 1.8, Equation 1.12 can be rewritten as

$$(1 - \phi_1 L - \phi_2 L^2 - \ldots - \phi_p L^p) y_t = c + \varepsilon_t \ . \tag{1.13}$$

It can be shown that such an AR(p)–process is stationary if all roots z_0 of the polynomial:

$$\phi_p(z) = 1 - \phi_1 z - \phi_2 z^2 - \ldots - \phi_p z^p \tag{1.14}$$

have modulus greater than one. The modulus of a complex number $z = z_1 + i z_2$ is defined as $|z| = \sqrt{z_1^2 + z_2^2}$. Viewing the stationarity condition from that point, it turns out that in the case of an AR(1)–process, like in Equation 1.8, $|\phi| < 1$ is required because the only solution to $1 - \phi z = 0$ is given for $z = 1/\phi$ and $|z| = |1/\phi| > 1$ when $|\phi| < 1$.

If the error process $\{\varepsilon_t\}$ is normally distributed, Equation 1.12 can be consistently estimated by the *ordinary least-squares* method (OLS). Furthermore, the OLS estimator for the unknown coefficient vector $\boldsymbol{\beta} = (c, \boldsymbol{\phi})'$ is asymptotically normal.

In Figure 1.4, an AR(2)–process is displayed and generated by

$$\hat{y}_t = 0.6 \hat{y}_{t-1} - 0.28 \hat{y}_{t-2} + \hat{\varepsilon}_t \ . \tag{1.15}$$

The *stability* of such a process can easily be checked with the function `poly-`

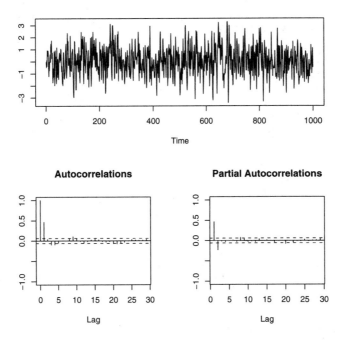

Fig. 1.4. Time series plot of AR(2)–process, $\phi_1 = 0.6$ and $\phi_2 = -0.28$

`root()`. In Rcode example 1.2, this AR(2)–processes has been first generated by using the function `filter()` instead of `arima.sim()` as in Rcode example 1.1 (see command line 2). The advantage of the former compared with the latter is that unstable AR(p)–processes can also be generated. Next, the generated AR(2)–process is estimated with `arima()`. The estimates are close to their theoretical values, as could be expected by a sample size of 1000. The moduli of the characteristic polynomial are retrieved with `Mod()`, the real and complex parts with the functions `Re()` and `Im()`, respectively. Please note that the signs of the estimated coefficients have to be reversed for the calculation of the roots (see command lines 7 and 8). The roots can be depicted in a cartesian coordinate system with a *unit circle*, as is shown in the next exhibit.

Rcode 1.2 Estimation of an AR(2)–process with $\phi_1 = 0.6$ and $\phi_2 = -0.28$

```
series <-  rnorm(1000)                                              1
y.st <- filter(series, filter=c(0.6, -0.28), method='          2
    recursive')
ar2.st <- arima(y.st, c(2, 0, 0), include.mean=FALSE,          3
    transform.pars=FALSE, method="ML")
ar2.st$coef                                                        4
polyroot(c(1, -ar2.st$coef))                                     5
Mod(polyroot(c(1, -ar2.st$coef)))                               6
root.comp <- Im(polyroot(c(1, -ar2.st$coef)))                  7
root.real <- Re(polyroot(c(1, -ar2.st$coef)))                  8
# Plotting the roots in a unit circle                             9
x <- seq(-1, 1, length = 1000)                                   10
y1 <- sqrt(1- x^2)                                                11
y2 <- -sqrt(1- x^2)                                               12
plot(c(x, x), c(y1, y2), xlab='Real part', ylab='Complex      13
    part', type='l', main='Unit Circle', ylim=c(-2, 2), xlim
    =c(-2, 2))
abline(h=0)                                                       14
abline(v=0)                                                       15
points(Re(polyroot(c(1, -ar2.st$coef))), Im(polyroot(c(1, -  16
    ar2.st$coef))), pch=19)
legend(-1.5, -1.5, legend="Roots of AR(2)", pch=19)          17
```

1.3 MA(q) Time Series Process

It has been shown in Section 1.2 that a finite stable AR(p)–process can be inverted to a moving average of contemporaneous and past shocks (see Equations 1.10a and 1.10b). We consider now how a process can be modeled as a finite *moving average* of its shocks. Such a process is called MA(q), where the parameter q refers to the highest lag of shocks to be included in such a process. We do so by analyzing an MA(1)–process first:

$$y_t = \mu + \varepsilon_t + \theta\varepsilon_{t-1} , \tag{1.16}$$

where $\{\varepsilon_t\}$ is a white noise process and μ, θ can be any constants. The moments of this MA(1)–process are given by,

$$\mu = E[y_t] = E[\mu + \varepsilon_t + \theta\varepsilon_{t-1}] , \tag{1.17a}$$

$$\gamma_0 = E[(y_t - \mu)^2] = (1 + \theta^2)\sigma^2 , \tag{1.17b}$$

$$\gamma_1 = E[(y_t - \mu)(y_{t-1} - \mu)] = \theta\sigma^2 . \tag{1.17c}$$

It is left to the reader to show that the higher autocovariances γ_j with $j > 1$ are nil. Neither the mean nor the autocovariance are functions of time, and

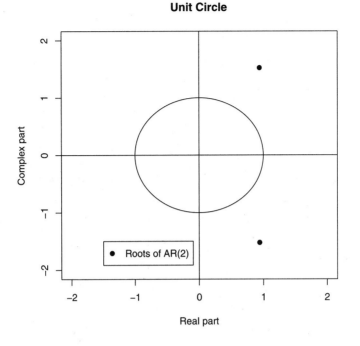

Fig. 1.5. Unit circle and roots of a stable AR(2)–process, $\phi_1 = 0.6$ and $\phi_2 = -0.28$

hence an MA(1)–process is covariance stationary for all values of θ. Incidentally, because Equation 1.6 is satisfied, this process also has the characteristic of ergodicity.

Likewise to the Rcode example 1.1, an MA(1)–process has been generated with $\mu = 0$ and $\theta = 0.8$ and is displayed in Figure 1.6. Let us now extend the MA(1)–process to the general class of MA(q)–processes:

$$y_t = \mu + \varepsilon_t + \theta_1 \varepsilon_{t-1} + \ldots + \theta_q \varepsilon_{t-q} . \tag{1.18}$$

With the lag operator L, this process can be rewritten as,

$$y_t - \mu = \varepsilon_t + \theta_1 \varepsilon_{t-1} + \ldots \theta_q \varepsilon_{t-q} \tag{1.19a}$$
$$= (1 + \theta_1 L + \ldots + \theta_q L^q)\varepsilon_t = \theta_q(L)\varepsilon_t . \tag{1.19b}$$

Similar to the case in which a stable AR(p)–process can be rewritten as an infinite MA–process, an MA(q)–process can be transformed to an infinite AR–process as long as the roots of the characteristic polynomial, the z-transform, have modulus greater than one, *i.e.*, are outside the unit circle:

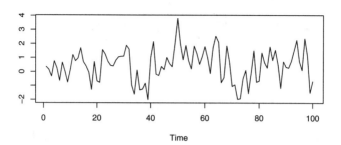

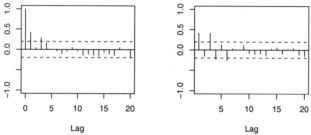

Fig. 1.6. Time series plot of MA(1)–process, $\theta = 0.8$

$$\theta_q z = 1 + \theta_1 z + \ldots + \theta_q z^q \; . \tag{1.20}$$

The expected value of an MA(q)–process is μ and hence invariant with respect to its order. The second-order moments are given as

$$\gamma_0 = E[(y_t - \mu)^2] = (1 + \theta_1^2 + \ldots + \theta_q^2)\sigma^2 \; , \tag{1.21a}$$

$$\gamma_j = E[(\varepsilon_t + \theta_1 \varepsilon_{t-1} + \ldots + \theta_q \varepsilon_{t-q})$$
$$\times \, (\varepsilon_{t-q} + \theta_1 \varepsilon_{t-j-1} + \ldots + \theta_q \varepsilon_{t-j-q})] \; . \tag{1.21b}$$

Because $\{\varepsilon_t\}$ are uncorrelated with each other by assumption, Equation 1.21b can be simplified to

$$\gamma_j = \begin{cases} (1 + \theta_{j+1}\theta_1 + \theta_{j+2}\theta_2 + \ldots + \theta_q\theta_{q-j})\sigma^2 & \text{for } j = 1, 2, \ldots, q \\ 0 & \text{for } j > q \; . \end{cases} \tag{1.22}$$

That is, empirically an MA(q)–process can be detected by its first q significant autocorrelations and a slowly decaying or alternating pattern of its partial autocorrelations. For large sample sizes T, a 95% significance band can be calculated as

$$\left(\varrho_j - \frac{2}{\sqrt{T}} , \; \varrho_j + \frac{2}{\sqrt{T}} \right) . \tag{1.23}$$

where ϱ_j refers to the jth-order autocorrelation.

It has been shown in Equation 1.10 that a finite AR–process can be inverted to an infinite MA–process. Before we proceed further, let us first examine the stability condition of such an MA(∞)–process:

$$y_t = \mu + \sum_{j=0}^{\infty} \psi_j \varepsilon_{t-j} . \tag{1.24}$$

Now, we acsribe the coefficients for an infinite process as ψ instead of θ, which was the case for MA(q)–processes. It can be shown that such an infinite process is covariance-stationary if the coefficient sequence $\{\psi_j\}$ is either square summable:

$$\sum_{j=0}^{\infty} \psi_j^2 < \infty \tag{1.25}$$

or absolute summable:

$$\sum_{J=0}^{\infty} |\psi_j| < \infty , \tag{1.26}$$

whereas absolute summability is sufficient for square summability, *i.e.*, the former implies the latter, but not *vice versa*.

1.4 ARMA(p, q) Time Series Process

It has been shown in the last two sections how a time series can be explained either by its history or by contemporaneous and past shocks. Furthermore, the moments of these data-generating processes have been derived and the mutual invertibility of these model classes have been stated for parameter sets that fullfill the stability condition. In this section, these two time series processes are put together, hence, a more general class of ARMA(p, q)–processes is investigated.

In practice, it is often cumbersome to detect a pure AR(p)– or MA(q)–process by the behavior of its empirical autocorrelation and partial autocorrelation functions, because neither one tapers off with increasing lag order. In these instances, the time series might have been generated by a mixed *autoregressive-moving average process*.
For a stationary time series $\{y_t\}$, such a mixed process is defined as

$$y_t = c + \phi_1 y_{t-1} + \ldots + \phi_p y_{t-p} + \varepsilon_t + \theta_1 \varepsilon_{t-1} + \ldots \theta_q \varepsilon_{t-q} . \tag{1.27}$$

By assumption, $\{y_t\}$ is stationary, *i.e.*, the roots of the characteristic polynomial lie outside the unit circle. Hence, with the lag operator, Equation 1.27 can be transformed to

$$y_t = \frac{c}{1 - \phi_1 L - \ldots - \phi_p L^p} + \frac{1 + \theta_1 L + \ldots + \theta_q L^q}{1 - \phi_1 L - \ldots - \phi_p L^p} \varepsilon_t \qquad (1.28a)$$

$$= \mu + \psi(L)\varepsilon_t . \qquad (1.28b)$$

The stated condition of absolute summability for the lag coefficients $\{\psi_j\}$ in Equation 1.26 must hold. Put differently, the stationarity condition depends only on the AR–parameters and not on the moving average ones.

Lastly, we will briefly touch on the *Box–Jenkins* [8] approach for time series modeling. This approach consists of three stages: identification, estimation, and diagnostic checking. As a first step, the series is visually inspected for stationarity. If an investigator has doubts that this condition is met, he or she has to suitably transform the series before proceeding. As we will see in the next chapter, such transformations could be the removal of a deterministic trend or taking first differences with respect to time. Furthermore, variance instability such as higher fluctuations as time proceeds can be coped with by using the logarithmic values of the series instead. By inspection of the empirical autocorrelation and partial autocorrelation functions, a temptative ARMA(p, q)–model is specified. The next stage is the estimation of the preliminary model. Here, one should check the model's stability as well as the significance of its parameters. If one of these tests fail, the econometrician has to start anew by specifying a more parsimonous model with respect to the ARMA–order. Now, let us assume that this is not the case. In the last step, diagnostic checking, he or she should then examine the residuals for uncorrelatedness, and normality and conduct tests for correctness of the model's order, *i.e.*, over- and underfitting. Incidentally, by calculating *pseudo-ex ante* forecasts, the model's suitability for prediction can be examined.

As an example, we will apply the Box–Jenkins approach to the unemployment rate of the United States (see Figure 1.2).[5] Because no trending behavior is visible, we first examine its autocorrelations functions (see command lines 7 and 8). The graphs are displayed in Figure 1.7. The autocorrelation function tapers off, whereas the partial autocorrelation function has two significant correlations. As a temptative order an ARMA(2, 0)–model is specified (see command line 11). This model is estimated with the `arima()` function in the `stats` package.

---

[5] We used the logarithmic values of the unemployment rate because of a changing variance with respect to time.

Rcode 1.3 Box–Jenkins: U.S. unemployment rate

```
library(urca)                                                    1
data(npext)                                                      2
y <- ts(na.omit(npext$unemploy), start=1909, end=1988,          3
   frequency=1)
op <- par(no.readonly=TRUE)                                      4
layout(matrix(c(1, 1, 2, 3), 2, 2, byrow=TRUE))                 5
plot(y, ylab="unemployment rate (logarithm)")                   6
acf(y, main='Autocorrelations', ylab='', ylim=c(-1, 1))         7
pacf(y, main='Partial Autocorrelations', ylab='', ylim=c       8
   (-1, 1))
par(op)                                                          9
# temptative ARMA(2,0)                                          10
arma20 <- arima(y, order=c(2, 0, 0))                            11
# Storing the residuals for diagnostic checking                12
res20 <- residuals(arma20)                                      13
tsdiag(arma20)                                                  14
# Diagnostic checking                                          15
# uncorrelatednes                                              16
Box.test(res20, lag = 20, type = "Ljung-Box")                  17
# normality                                                    18
shapiro.test(res20)                                            19
# overfitting                                                 20
arma30 <- arima(y, order=c(3, 0, 0))                           21
```

The values of the estimated AR coefficients are $\phi_1 = 0.9297$ and $\phi_2 =$ -0.2356. Their estimated standard errors are 0.1079 and 0.1077. Both AR coefficients are significantly different from zero, and the estimated values satisfy the stability condition. In the next step, the model's residuals are retrieved and stored in the object res20. Likewise to the unemployment series, the residuals can be inspected visually as well as their autocorrelation function (ACF) and partial autocorrelation function (PACF). Furthermore, the assumption of uncorrelatedness can be tested with the *Ljung–Box Portmanteau test* [57]. This test is implemented in the Box.test() function in the stats package. Except for the PACF, these tools are graphically returned by the function tsdiag() (see Figure 1.8). The null hypothesis of uncorrelatedness up to order 20 cannot be rejected, given a marginal significance level of 0.3452. The hypothesis of normally distributed errors can be tested with the *Jarque–Bera test* for normality [5][6]: jarque.bera.test() contained in the contributed CRAN package tseries by Trapletti and Hornik [96] or with the *Shapiro–Wilk test* [91][92]: shapiro.test(), for example. Given a p-value of 0.9501, the normality hypothesis cannot be rejected. It should be noted that the assumption of normality could be visually inspected by a normal *quantile plot* (qqnorm()). As a last diagnostic test an overparameterized ARMA(3,

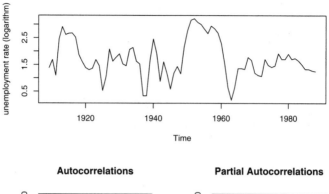

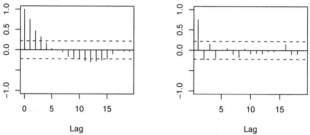

Fig. 1.7. Time series plot, ACF, and PACF of U.S. unemployment rate

0)–model is estimated. It turns out that first, the coefficient for the third lag is not significantly different from zero, and that second, the estimates for the first- and second-order AR–coefficients remain almost unchanged. Although the model's fit could be improved by including dummy variables to take into account the wide swings of the series during the pre-World War II era, but by now we conclude that the U.S. unemployment rate can be well represented by an ARMA(2, 0)–model.

Summary

In this first chapter the definition of a time series and the concept of its data-generating process has been introduced. You should now be familiar with how to characterize a time series by its moments and distinguish the different concepts of stationarity. Two model classes for a time series also have been introduced, namely the autoregressive and the moving average model, as well as a combination thereof. You should be able to detect and distinguish the order of these models by investigation of its autocorrelation and partial autocorrelation function. Finally, the Box–Jenkins approach to time series analysis has

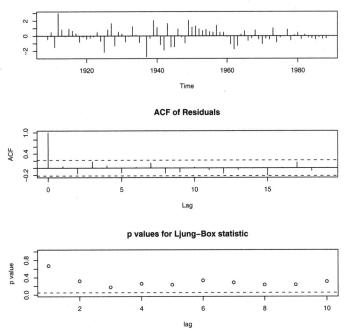

Fig. 1.8. Time series plot, ACF, and Ljung-Box of residuals [ARMA(2, 0)]

been presented. It is decomposed into three stages – specification of a temptative model order, estimation, and diagnostic checking.

So far, we have restricted the presentation to stationary time series only. At first sight this focus might seem to be too myopic, given that many time series cannot be characterized by a stationary process, in particular in macroeconomic and financial data sets. Therefore in the next chapter, nonstationary time series processes are discussed, including how these can be transformed to achieve stationarity.

Exercises

1. Derive the second-order moments of an AR(1)–process as in Equation 1.8.
2. Generate stable AR(1)–, AR(2)–, and AR(3)–processes with $T = 1000$ for different error variances, and plot their autocorrelations and partial autocorrelations. How could you determine the order of an AR(p)–process from its sample moments?
3. Show that the autocovariances $j > q$ of an MA(q)–process are zero.
4. Generate stable MA(1)–, MA(2)–, and MA(3)–processes with $T = 1000$ for different error variances, and plot their autocorrelations and partial autocorrelations. How could you determine the order of an MA(p)–process from its sample moments?

2

Nonstationary Time Series

In this chapter, models for nonstationary time series are introduced. Before the characteristics of unit processes are presented, the differences between trend- and difference-stationary models are outlined. In the last section, long memory processes, i.e., fractionally integrated processes, are presented as a bridge between stationary and unit root processes.

2.1 Trend- *versus* Difference-Stationary Series

In the last chapter, a model class for stationary time series has been introduced. For instance, it has been shown that a stable autoregressive (AR)(p) can be inverted to an infinite moving average (MA)–process with a constant mean. However, most macroeconomic time series seem not to adhere to such a data-generating process (see Figure 1.1). In this section, we will for examplary reasons consider a more encompassing data-generated process that was presented by Campbell and Perron [9].

Now, it is assumed that a time series $\{y_t\}$ is a realization of a deterministic trend and a stochastic component:

$$y_t = TD_t + z_t \, , \tag{2.1}$$

where TD_t assigns a deterministic trend: $TD_t = \beta_1 + \beta_2 t$ and z_t represents the stochastic component: $\phi(L)z_t = \theta(L)\varepsilon_t$ with $\varepsilon_t \sim$ i.i.d, *i.e.*, an autoregressive-moving average (ARMA)(p, q)–process. We distinguish two cases. First, if all roots of the autoregressive polynomial lie outside the unit circle (see Equation 1.14), then $\{y_t\}$ is stationary around a deterministic trend. In this instance, one could remove the trend from the original series $\{y_t\}$ and fit an ARMA(p, q) to the residuals.[1]

This *trend-stationary* model is also termed an integrated model of order zero or more compactly, the $I(0)$–model. Second, assume now that one root of the autoregressive polynomial lies on the unit circle and the remaining ones are all outside. Here, $\Delta z_t = (1 - L)z_t$ is stationary around a constant mean.

---

[1] A deterministic trend is most easily subtracted from a series, *i.e.*, a vector y, by issuing the following command: `detrended <- residuals(lm(y ~ seq(along=y)))`.

The series $\{y_t\}$ is *difference-stationary* because one has to apply the first difference filter with respect to time to obtain a stationary process. Likewise to the trend-stationary model this difference stationary model is referred to as an integrated model of order one or shortly the $I(1)$–model. The meaning of "intergrated" should now be obvious: Once the series has been differenced to obtain a stationary process, it must be integrated once, *i.e*, the reversal, to achieve the original series, hence the $I(1)$–model. An ARMA(p, q)–model could then be fitted to the differenced series. This model class is termed the *autoregressive integrated moving average* (ARIMA)(p, d, q), where d refers to the order of integration, *i.e.*, how many times the original series must be differenced until a stationary one is obtained. It should be noted that unit roots, *i.e.*, roots of the autoregressive polynomial that lie on the unit circle, are solely referring to the stochastic component in Equation 2.1.

The distinction between a trend- and a difference-stationary processes is examplified by the following two processes:

$$y_t = y_{t-1} + \mu = y_0 + \mu t \ , \tag{2.2a}$$

$$y_t = y_{t-1} + \varepsilon_t = y_0 + \sum_{s=1}^{t} \varepsilon_s \ , \tag{2.2b}$$

where μ is a fixed constant and ε_t is a white noise process. In Equation 2.2a, $\{y_t\}$ is represented by a deterministic trend, whereas in Equation 2.2b, the series is explained by its cumulated shocks, *i.e.*, a stochastic trend.

So far, the stochastic component z_t has been modeled as an ARIMA(p, d, q)–model. To foster the understanding of unit roots, we will decompose the stochastic component into a cyclical component c_t and a stochastic trend TS_t. It is assumed that the cyclical component is a mean-stationary process, whereas all random shocks are captured by the stochastic component. Now, the data-generating process for $\{y_t\}$ is decomposed into a *deterministic trend*, a *stochastic trend*, and a *cyclcical component*. For the trend stationary model, the stochastic trend is zero and the cyclical component is equal to the ARMA(p, q)–model: $\phi(L)z_t = \theta(l)\varepsilon_t$. In the case of a difference-stationary model, the autoregressive polynomial contains a unit root that can be factored out: $\phi(L) = (1 - L)\phi^*(L)$, whereby the roots of the polynomial $\phi^*(L)$ are outside the unit circle. It is then possible to express Δz_t as a moving average process (for comparison, see Equations 1.28a and 1.28b):

$$\phi^*(L)\Delta z_t = \theta(L)\varepsilon_t \ , \tag{2.3a}$$

$$\Delta z_t = \phi^*(L)\theta(L)\varepsilon_t \ , \tag{2.3b}$$

$$\Delta z_t = \psi(L)\varepsilon_t \ . \tag{2.3c}$$

Beveridge and Nelson have shown that Equation 2.3c can be transformed to (*Beveridge–Nelson decomposition*)

$$z_t = TS_t + c_t = \psi(1)S_t + \psi^*(L)\varepsilon_t \;, \tag{2.4}$$

where the sum of the moving average coefficients is denoted by $\psi(1)$, S_t is the sum of the past and present random shocks: $\sum_{s=1}^{t} \varepsilon_s$, and the polynomial $\psi^*(L)$ is equal to $(1-L)^{-1}[\psi(L) - \psi(1)]$ (see Beveridge and Nelson [7]).

The time series $\{y_t\}$ is now explained by a trend function that consists of a deterministic trend as well as a stochastic component, namely $TS_T = \psi(1)S_t$. The latter affects the absolute term in each period. Because the stochastic trend is defined as the sum of the moving average coefficients of Δz_t, it can be interpreted as the long-run impact of a shock to the level of z_t. In contrast, the cyclical component, $c_t = \psi^*(L)\varepsilon_t$ exerts no long-run impact on the level of z_t. Now, we can distinguish the following four cases: (1) $\psi(1) > 1$: The long-run impact of the shocks is greater than the intermediate ones, and hence the series is characterized by an explosive path; (2) $\psi(1) < 1$: The impact of the shocks diminishes as time passes by, (3) $\psi(1) = 0$: The time series $\{y_t\}$ is a trend-stationary process, and (4) $\psi(1) = 1$: The data-generated process is a random walk. The fourth case will be a subject in the next section.

2.2 Unit Root Processes

As stated in the last section, if the sum of the moving average coefficients, $\psi(1)$ equals one, a *random walk* process results. This data-generating process has attracted much interest in the empirical literature, in particular in the field of financial econometrics. Hence, a random walk is not only a prototype for a unit root process, but it is implied by economic and financial hypotheses as well (*i.e.*, the efficient market hypothesis). Therefore, we will begin this section by analyzing random walk processes in more detail before statistical tests and strategies for detecting unit roots are presented.

A pure random walk without a drift is defined as

$$y_t = y_{t-1} + \varepsilon_t = y_0 + \sum_{s=1}^{t} \varepsilon_t \;, \tag{2.5}$$

where $\{\varepsilon_t\}$ is an i.i.d. process, *i.e.*, white noise. For the sake of simplicity, assume that the expected value of y_0 is zero and that the white noise process $\{\varepsilon_t\}$ is independent of y_0. Then it is trivial to show that (1) $E[y_t] = 0$ and $var(y_t) = t\sigma^2$. Clearly, a random walk is a *nonstationary time series* process because its variance grows with time. Second, the best forecast of random walk is its value one period earlier, *i.e.*, $\Delta y_t = \varepsilon_t$. Incidentally, it should be noted that the i.i.d. assumption for the error process $\{\varepsilon_t\}$ is important with respect to the conclusions drawn above. Suppose that the data-generated process for $\{y_t\}$ is

$$y_t = y_{t-1} + \varepsilon_t \ , \ \varepsilon_t = \rho \varepsilon_{t-1} + \xi_t \ , \tag{2.6}$$

where $|\rho| < 1$ and ξ_t is a white noise process instead. Then, $\{y_t\}$ is not a random walk process, but it still has a unit root and is a first-order nonstationary process.

Let us now consider the case of a random walk with drift:

$$y_t = \mu + y_{t-1} + \varepsilon_t = y_0 + \mu t + \sum_{s=1}^{t} \varepsilon_t \ , \tag{2.7}$$

where, likewise to the pure random walk process, $\{\varepsilon_t\}$ is white noise. For $\mu \neq 0$, $\{y_t\}$ contains a deterministic trend with drift parameter μ. The sign of this drift parameter causes the series to wander upward if positive and downward if negative, whereas the size of the absolute value affects the steepness.

In Rcode example 2.1, three time series have been generated. For a better comparability between those, all series have been calculated with the same sequence of random numbers drawn from a standard normal distribution. First, a pure random walk has been generated by calculating the cumulated sum of 500 random numbers stored in the vector object **e**. A deterministic trend has been set with the short form of the **seq()** function, *i.e.*, the colon operator. As a second time series model, a random walk with drift can now be easily created acording to Equation 2.7. Last, the deterministic trend has been overlayed with the stationary series of normally distributed errors. All three series are plotted in Figure 2.1. By ocular econometrics, it should be evident that the statistical discrimination between a deterministic trend contaminated with noise and a random walk with drift is not easy. Likewise, it is difficult to distinguish between a random walk process and a stable AR(1)–process in which the autoregressive coefficient is close to unity. The latter two time series processes are displayed in Exhibit 2.2.

Rcode 2.1 Stochastic and deterministic trends

```
set.seed(123456)                                              1
e <- rnorm(500)                                               2
# pure random walk                                            3
rw.nd <- cumsum(e)                                            4
# trend                                                       5
trd <- 1:500                                                  6
# random walk with drift                                      7
rw.wd <- 0.5*trd + cumsum(e)                                  8
# deterministic trend and noise                               9
dt <- e + 0.5*trd                                            10
# plotting                                                   11
par(mar=rep(5,4))                                            12
plot.ts(dt, lty=1, ylab='', xlab='')                        13
lines(rw.wd, lty=2)                                          14
par(new=T)                                                   15
plot.ts(rw.nd, lty=3, axes=FALSE)                           16
axis(4, pretty(range(rw.nd)))                               17
lines(rw.nd, lty=3)                                         18
legend(10, 18.7, legend=c('det. trend + noise (ls)', 'rw    19
    drift (ls)', 'rw (rs)'), lty=c(1, 2, 3))
```

Before a testing procedure for the underlying data-generating process is outlined, we will introduce a formal defintion of integrated series and briefly touch on the concept of *seasonal integration*, which will be presented in more detail in Section 5.2.

In the seminal paper by Engle and Granger [26] an integrated series is defined as follows.

Definition 2.1. *A series with no deterministic component that has a stationary, invertible, ARMA representation after differencing* d *times is said to be integrated of order* d, *which is denoted as* $x_t \sim I(d)$.

That is, a stationary series is simply written as an $I(0)$–process, whereas a random walk is said to follow an $I(1)$–process, because it has to be differenced once, before stationarity is achieved. It should be noted at this point that some macroeconomic series are already differenced. For example, the real net investment in an economy is the difference of its capital stock. If investment is an $I(1)$–process, then the capital stock must behave like an $I(2)$–process. Similarly, if the inflation rate, measured as the difference of the logarithmic price index, is integrated of order one, then the price index follows an $I(2)$–process. Therefore, stationarity of $y_t \sim I(2)$ is achieved by taking the first differences of the first differences:

$$\Delta\Delta y_t = \Delta(y_t - y_{t-1}) = (y_t - y_{t-1}) - (y_{t-1} - y_{t-2}) = y_t - 2y_{t-1} + y_{t-2} \ . \quad (2.8)$$

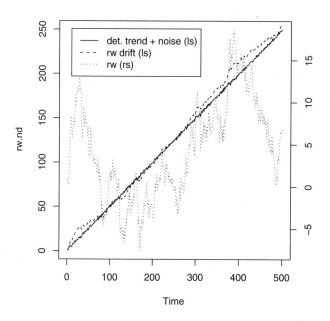

Fig. 2.1. Time series plot of deterministic and stochastic trends

If a series is already stationary $I(0)$, then no further differencing is necessary. When a series $\{y_t\}$ is a linear combination of $x_{1t} \sim I(0)$ and $x_{2t} \sim I(1)$, then $\{y_t\}$ will be an $I(1)$–process. Likewise, a linear transformation of an $I(d)$–process conserves the order of integration: $y_t \sim I(d)$, so it will be $\alpha + \beta y_t \sim I(d)$, where α and β are constants.

By now, we have only considered data-generating processes in which the unit root occurs for its own values lagged by one period. One can generalize these processes to

$$y_t = y_{t-s} + \varepsilon_t \,, \tag{2.9}$$

where $s \geq 1$. If s equals a seasonal frequency of the series, then $\{y_t\}$ is determined by its prior seasonal values plus noise. Likewise to the concept of a stochastic trend, this data-generating process is termed *stochastic seasonality*. In practice, seasonality is often accounted for by the inclusion of seasonal dummy variables or the use of seasonally adjusted data. However, there might be instances where the allowance of a seasonal component to drift over time is necessary. Analogously to the presentation of the unit root processes at the zero frequency, we can define the lag operator for seasonal unit roots as

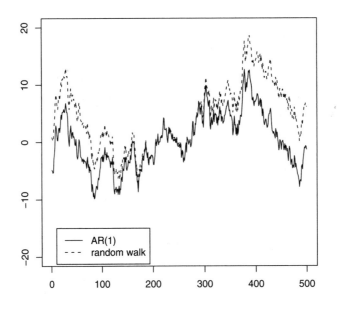

Fig. 2.2. Plot of a random walk and a stable AR(1)–process, $\phi = 0.99$

$$\Delta_s = (1 - L^s) \tag{2.10a}$$
$$= (1 - L)(1 + L + L^2 + \ldots + L^{s-1}) = \Delta S(L) \,. \tag{2.10b}$$

In Equation 2.10b the unit root at the zero frequency has been factored out. Hence, a seasonally integrated series can be represented as the product of the first difference operator and the moving-average seasonal filter $S(L)$. According to Engle *et al.* [27] a seasonally integrated series can be defined as follows.

Definition 2.2. *A variable $\{y_t\}$ is said to be seasonally integrated of orders* d *and* D*, which are denoted as* SI*(*d*,* D*), if* $\Delta^d S(L)^D y_t$ *is stationary.*

Therefore, if a quarterly series $\Delta_4 y_t$ is stationary, then $\{y_t\}$ is $SI(1,1)$. Testing for seasonal unit roots is similar although a bit more complicated to testing for unit roots at the zero freqeuncy, which will be presented in the next paragraphs. The probably simplest test has been proposed by Hasza and Fuller [40], Dickey *et al.* [18], and a modification of it by Osborn *et al.* [71]. However, more complicated testing procedure that allows for cyclical movements at different frequencies has been introduced into the literature by Hylleberg *et al.* [49]. In R, seasonal unit root tests are implemented in the CRAN–package uroot.

Recall the decomposition of a time series $\{y_t\}$ as in Equation 2.1. Now we want to investigate if the process $\{z_t\}$ does contain a unit root:

$$z_t = y_t - TD_t \, . \tag{2.11}$$

Hence, a deterministic trend is removed from the original series first and the residuals are tested for a unit root. Dickey and Fuller [16] proposed the following test regression that is delineated from an assumed AR(1)–process of $\{z_t\}$ (henceforth: DF–test):

$$z_t = \theta z_{t-1} + \varepsilon_t \, , \tag{2.12a}$$
$$z_t - z_{t-1} = \theta z_{t-1} - z_{t-1} + \varepsilon_t \, , \tag{2.12b}$$
$$\Delta z_t = (\theta - 1)z_{t-1} + \varepsilon_t \, , \tag{2.12c}$$
$$\Delta z_t = \pi z_{t-1} + \varepsilon_t \, . \tag{2.12d}$$

Under the null hypothesis of a unit root $\pi = 0$, which is equivalent to $\theta = 1$ and the alternative is a trend stationary process, i.e., $\pi < 0$ or $\theta < 1$. Please note, that an explosive path for $\{z_t\}$, $\pi > 0$, is excluded. Equation 2.12d can be estimated by the ordinary least-squares method. The significance of π can be tested by a usual Student t ratio. However, this test statistic does not have the familiar Student t distribution. Under the null hypothesis, an $I(0)$–variable is regressed on an $I(1)$–variable in Equation 2.12d. In this case, the limiting distribution of the Student t ratio is not normal. Fortunately, critical values have been calculated by simulation and are publicized in Fuller [32], Dickey and Fuller [17], and MacKinnon [62], for instance.

So far we have only stated that a deterministic trend is removed before testing for a unit root. In reality neither the existence nor the form of the deterministic component is known a priori. Hence, we have to choose from the set of deterministic variables DV_t the one that best suites the data-generating process. The most obvious candidates as DV_t are simply a constant, a linear trend, or higher polynomials in the trend function, i.e., square or cubic. In practice, only the first two are considered. The aim of characterizing the noise-function $\{z_t\}$ is still the same, but now we have to take the various DV_t as deterministic regressors DR_t into account too. The above-described two-step procedure (Equations 2.11 and 2.12) can be carried out in one equation:

$$\Delta y_t = \boldsymbol{\tau}'DR_t + \pi y_{t-1} + u_t \, , \tag{2.13}$$

where $\boldsymbol{\tau}$ is the coefficient vector of the deterministic part and $\{u_t\}$ assigns an error term. For the one-step procedure, a difficulty now arises, because under the validity of the null hypothesis, the deterministic trend coefficient $\boldsymbol{\tau}$ is null, whereas under the alternative it is not. Hence, the distribution of the Student t ratio of π depends now on these nuisance parameters too. The reason for this is that the true deterministic component is unknown and must be estimated. Critical values for different deterministic components can be found in the above-cited literature as well as in Ouliaris *et al.* [72].

A weakness of the original DF–test is that it does not take a possible serial correlation of the error process $\{u_t\}$ into account. Dickey and Fuller [17] have suggested replacing the AR(1)–process for $\{z_t\}$ in Equation 2.12a with an ARMA(p, q)–process, $\phi(L)z_t = \theta(L)\varepsilon_t$. If the noise component is an AR(p)–process, it can be shown that the following test regression:

$$\Delta y_t = \tau' DR_t + \pi y_{t-1} + \sum_{j=1}^{k} \gamma_j \Delta y_{t-j} + u_t \text{ with } k = p - 1 \qquad (2.14)$$

ensures that the serial correlation in the error is removed. This test regression is called the *augmented Dickey–Fuller* test, henceforth, the ADF–test. Several methods for selecting k have been suggested in the literature. The most prominent one is the *general-to-specific* method. Here, one starts with an *a priori* choosen upper bound k_{max} and then drops the last lagged regressor if it is insignificant. In this case, the Student t distribution is applicable. You repeat these steps until the last lagged regressors is significant, otherwise you drop it each time the equation is reestimated. If no endogenously lagged regressor turns out to be significant, you choose $k = 0$, hence the DF–test results. This procedure will asymptotically yield the correct lag order or greater to the true order with probability one. Other methods for selecting an appropriate order k are based on information criteria, like Akaike (AIC) [1] or Schwarz (SC) [90]. Alternatively, the lag order can be determined by testing the residuals for a lack of serial correlation as can be tested *via* the Ljung–Box Portmanteau test (LB) or a Lagrange multiplier test (LM). In general, the SC, LB, or LM tests coincide with respect to selecting an optimal lag length k. Whereas the AIC and the general-to-specific method will mostly imply a lag length at least as large as the one by the former methods.

Once the lag order k is empirically determined, the next steps involve a *testing procedure* as graphically illustrated in Figure 2.3. First, the encompassing ADF–test equation:

$$\Delta y_t = \beta_1 + \beta_2 t + \pi y_{t-1} + \sum_{j=1}^{k} \gamma_j \Delta y_{t-j} + u_t \qquad (2.15)$$

is estimated. Dependent on this result are the further steps to be taken, until one can conclude that the series is

  i) stationary around a zero mean,
 ii) stationary around a nonzero mean,
iii) stationary around a linear trend,

iv) contains a unit root with zero drift,
v) contains a unit root with nonzero drift.

To be more concrete, the testing strategy starts by testing if $\pi = 0$ using the t statistic τ_τ. This statistic is not standard Student t distributed, but critical values can be found in Fuller [32]. If this test is rejected, then there is no need to proceed further. The testing sequence is continued by an F type test Φ_3 with $H_0 : \beta_2 = \pi = 0$ using the critical values tabulated in Dickey and Fuller [17]. If it is significant, then test again for a unit root using the standardized normal. Otherwise if the hypothesis $\beta_2 = 0$ cannot be rejected, reestimate the Equation 2.15 but without a trend. The corresponding t and F statistics for testing if $H_0 : \pi = 0$ and $H_0 : \beta_1 = \pi = 0$ are denoted by $\tau_\mu(\tau)$ and Φ_1. Again, the critical values for these test statistics are provided in the above-cited literature. If the null hypothesis of $\tau_\mu(\tau)$ is rejected, then there is again no need to go further. If it is not, then employ the F statistic Φ_1 for testing of the presence of a constant and a unit root.

However, the testing procedure does not end here. If the hypothesis $\pi = 0$ cannot be rejected in Equation 2.15, then the series might be integrated of order higher than zero. Therefore, one has to test whether the series is $I(1)$ or possibly $I(2)$ or even integrated to a higher degree. A natural approach would be to apply the DF– or ADF–test to

$$\Delta\Delta y_t = \pi \Delta y_{t-1} + u_t . \tag{2.16}$$

If the null hypothesis $\pi = 0$ is rejected, then $\Delta y_t \sim I(0)$ and $y_t \sim I(1)$, otherwise one must test subsequently whether $y_t \sim I(2)$. This testing procedure is termed *bottom-up*. However, two possibilities arise by using this bottom-up approach. First, the series cannot be transformed to stationarity regardless of how many times the difference operator is applied. Second, the danger of overdifferencing exists; that is, one falsly concludes an integration order higher than the true one. This can be detected by high positive values of the DF–test statistic. This risk can be circumvented by a general-to-specific testing strategy proposed by Dickey and Pantula [19]. They recommend by starting from the highest sensible order of integration, say $I(2)$, and then test downward to the stationary case.

So far, we have only considered the DF– and the ADF–test as a means to detect the presence of unit roots. Since the early 1980s numerous other statistical tests have been proposed in the literature. The most important and widely used ones will be presented in the second part of the book.

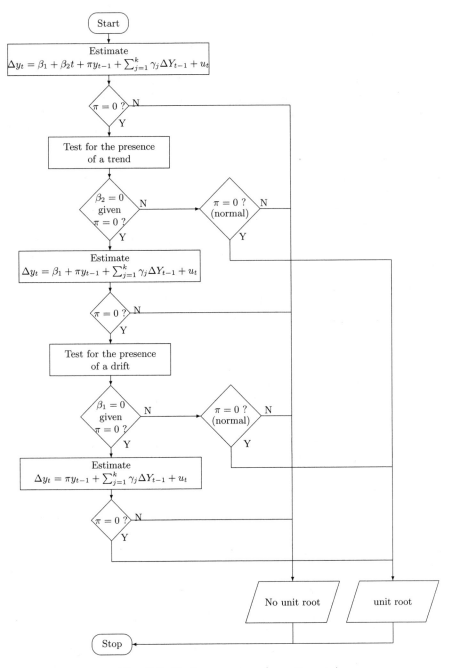

Fig. 2.3. Testing sequence for unit roots

2.3 Long Memory Processes

So far, we have considered data-generating processes that are either stationary or integrated of an integer order higher than zero (for example, the random walk as a prototype of an $I(1)$ series). Hence, it is a knife-edge decision if a series is $I(1)$ or $I(0)$ or is integrated at an even higher integer order. Furthermore, it has been shown that for a $y_t \sim I(1)$ series, the ACF declines linearly, and for a stationary $y_t \sim I(0)$ process, the ACF declines exponentially so that observations separated by a long time span may be regarded as independent. However, some empirically observed time series do not share neither one of these characteristics even though they are transformed to stationarity by suitable differencing. These time series still exhibit a dependency between distant observations. Their occurences are encountered in many disciplines such as finance, geophysical sciences, hydrology, and marcoeconomics. Although having argued heuristically, Granger [34] provides a theoretical justification for these processes. To cope with such time series, our current model class has to be enlarged by so-called *fractionally integrated* processes, *i.e., long memory processes*. The literature about fractionally integrated processes has grown steadily since its detection in the early 1950s of the last century. Baillie [4] cites in his survey about these processes 138 articles and 38 background references.

Before the more encompassing class of *autoregressive fractionally integrated moving average* processes (henceforth: ARFIMA) is introduced, it is noteworthy to define a long memory process and the filter for transforming fractionally integrated series.

First, we draw on the definition of McLeod and Hipel [67].

Definition 2.3. *A process is said to posses a long memory if*

$$\lim_{T \to \infty} \sum_{j=-T}^{T} |\rho_j| \tag{2.17}$$

is nonfinite.

This is equivalent by stating that the spectral density of a long memory process becomes unbounded at low frequencies.[2]

Second, recall that an integrated process of order d can be written as

---

[2] For an exposition of frequency domain analysis, the reader is referred to Judge, G. G., W. E. Griffiths, R. C. Hill, H. Lütkepohl, and T. Lee, The Theory and Practice of Econometrics, John Wiley and Sons, New York, 2nd edition, 1985, and Bloomfield, P., Fourier Analysis of Time Series: An Introduction, John Wiley and Sons, New York, 2nd edition, 2000. The spectral density of a series can be estimated by the function `spectrum()` in R. For more information on how this is implemented, the reader is referred to Venables and Ripley [98] as well as the function's documentation.

$$(1 - L)^d y_t = \psi(L)\varepsilon_t \ , \tag{2.18}$$

where absolute or square summabilty of ψ_j is given; *i.e.*, $\sum_{j=0}^{\infty} |\psi_j| < \infty$ or $\sum_{j=0}^{\infty} \psi_j^2 < \infty$. Premultiplying Equation 2.18 by $(1 - L)^{-d}$ yields

$$y_t = (1 - L)^{-d}\psi(L)\varepsilon_t \ . \tag{2.19}$$

Now, define the function $f(z) = (1 - z)^{-d}$ for the scalar z. The derivatives of this function are

$$\frac{\mathrm{d}f}{\mathrm{d}z} = d(1 - z)^{-d-1} \ , \tag{2.20a}$$

$$\frac{\mathrm{d}^2 f}{\mathrm{d}z^2} = (d + 1)d(1 - z)^{-d-2} \ , \tag{2.20b}$$

$$\vdots$$

$$\frac{\mathrm{d}^j f}{\mathrm{d}z^j} = (d + j - 1)(d + j - 2) \cdots (d + 1)d(1 - z)^{-d-j} \ . \tag{2.20c}$$

Therefore, the fractional difference operator for $d \in (-\frac{1}{2}, \frac{1}{2}]$ can be expressed as

$$(1 - L)^d = \sum_{j=0}^{\infty} \binom{d}{j}(-1)^j L^j \ , \tag{2.21}$$

by making use of a power series expansion around $z = 0$ and the binomial theorem. The coefficient sequence $\binom{d}{j}(-1)^j$ is square summable and can be expressed in terms of the gamma function $\Gamma()$ as

$$\binom{d}{j}(-1)^j = \frac{\Gamma(-d + j)}{\Gamma(-d)\Gamma(j + 1)} \ . \tag{2.22}$$

Two points are worthy to note. When $d > \frac{1}{2}$, an integer difference operator can be applied first. Incidentally, in this case, the process becomes nonstationary with unbounded variance. Robinson [85] calls such a process "less nonstationary" than a unit root process, smoothly bridging the gulf between $I(0)$– and $I(1)$–processes. Second, because in pratice no series with infinite observations are at hand, one truncates the expression in Equation 2.21 for values y_{t-j} outside the sample range and sets $y_{t-j} = 0$:

$$y_t^* = \sum_{j=0}^{\infty} \frac{\Gamma(-d + j)}{\Gamma(-d)\Gamma(j + 1)} \, y_{t-j} \ , \tag{2.23}$$

where y_t^* assigns the fractional differenced series.

The now to be introduced ARFIMA(p, d, q) class has been developed independently by Granger and Joyeux [36] and Hosking [47]. The estimation and simulation of these models is implemented in R within the contributed package `fracdiff`. Formally, an ARFIMA(p, d, q)–model is defined as follows.

Definition 2.4. *The series $\{y_t\}$ is an invertible and stationary ARFIMA(p, d, q)–process if it can be written as*

$$\Delta^d y_t = z_t , \qquad (2.24)$$

where $\{z_t\}_{t=-\infty}^{\infty}$ is an ARMA(p, q)–process such that $z_t = \phi_p(L)^{-1}\theta_q(L)\varepsilon_t$ and both lag polynomials have their roots outside the unit circle, where ε_t is a zero-mean i.i.d. random variable with variance σ^2 and $d \in (-0.5, 0.5]$.

For parameter values $0 < d < 0.5$, the process is long memory, and for the range $-0.5 < d < 0$, the sum of absolute values of its autocorrelations tends to a constant. In this case, the process exhibits negative dependency between distant observations and is therefore termed "*antipersistent*" or to have "*intermediate memory.*" Regardless, whether the process $\{y_t\}$ is long memory or intermediate memory, as long as $d > -0.5$, it has an invertible moving average representation. How is the long memory behavior incorporated in such a process? It can be shown that the autocorrelation function (ACF) of long memory processes declines hyperbolically instead of exponentially as would be the case for stable ARMA(p, q)–models. The speed of the decay depends on the parameter value d. For instance, given a fractional white noise process ARFIMA(0, d, 0), Granger and Joyeux [36] and Hosking [47] have proven that the autocorrelations are given by

$$\rho_j = \frac{\Gamma(j+d)\Gamma(1-d)}{\Gamma(j-d+1)\Gamma(d)} . \qquad (2.25)$$

The counterpart of this behavior in the frequency domain analysis is an unbounded spectral density as the frequency ω tends to zero. In the Rcode example 2.2, an ARIMA(0.4, 0.0, 0.0) and an ARFIMA(0.4, 0.4, 0.0) have been generated and their ACFs as well as spectral densities are displayed in Figure 2.4.

Rcode 2.2 ARMA versus ARFIMA model

```
library(fracdiff)                                                          1
set.seed(123456)                                                           2
# ARFIMA(0.4,0.4,0.0)                                                      3
y1 <- fracdiff.sim(n=1000, ar=0.4, ma=0.0, d=0.4)                         4
# ARIMA(0.4,0.0,0.0)                                                       5
y2 <- arima.sim(model=list(ar=0.4), n=1000)                               6
# Graphics                                                                 7
op <- par(no.readonly=TRUE)                                                8
layout(matrix(1:6, 3, 2, byrow=FALSE))                                     9
plot.ts(y1$series, main='Time series plot of long memory',               10
    ylab='')
acf(y1$series, lag.max=100, main='Autocorrelations of long              11
    memory')
spectrum(y1$series, main='Spectral density of long memory')             12
plot.ts(y2, main='Time series plot of short memory', ylab=''             13
    )
acf(y2, lag.max=100, main='Autocorrelations of short memory'            14
    )
spectrum(y2, main='Spectral density of short memory')                    15
par(op)                                                                   16
```

A long memory series with 1000 observations has been generated with the function `fracdiff.sim()` contained in the package `fracdiff`, whereas the short memory series has been calculated with the function `arima.sim()` (see command lines 4 and 6).[3] As can be clearly seen in Figure 2.4, the autocorrelations decline much more slowly compared with the stationary AR(1)–model and its spectral density is as $\omega \to 0$ higher about a factor of 100.

By now, the question of how to estimate the fractional difference parameter d or to detect the presence of long memory behavior in a time series is unanswered. We will now present three approaches to do so, whereas the last one deals with the simultaneous estimation of all parameters in an ARFIMA(p, d, q)–model.

The classic approach for detecting the presence of long-term memory can be found in Hurst [48]. He proposed the rescaled range statistic, or shortly the R/S statistic. This descriptive measure is defined as

$$R/S = \frac{1}{s_T}\left[\max_{1\le k\le T}\sum_{j=1}^{k}(y_j - \bar{y}) - \min_{1\le k\le T}\sum_{j=1}^{k}(y_j - \bar{y})\right], \qquad (2.26)$$

---

[3] Functions for generating and modeling long memory series can also be found in the contributed CRAN package `fSeries` [102].

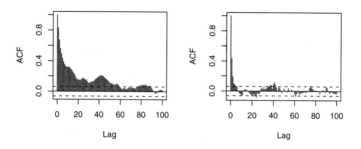

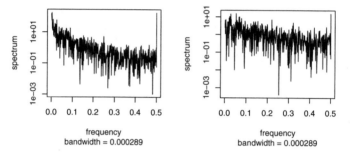

Fig. 2.4. Graphical display: ARIMA versus ARFIMA

where s_T is the usual maximum likelihood standard deviation estimator: $s_T = [\frac{1}{T}\sum_{j=1}^{T}(y_j-\bar{y})^2]^{\frac{1}{2}}$. This measure is always nonnegative because the deviations from the sample mean $\bar{y}$ sum up to zero. Hence, the maximum of the partial sums will always be positive and likewise the minimum will always be negative. Hurst [48] showed the probability limit of

$$\plim_{T\to\infty}\left\{T^{-H}(\frac{R/S}{s_t})\right\} = \text{const} ,\tag{2.27}$$

where H assigns the *Hurst coefficient*. The Hurst coefficient is then estimated as

$$\hat{H} = \frac{\log(R/S)}{\log(T)} .\tag{2.28}$$

A short memory process is associated with a value of $H = \frac{1}{2}$, and estimated values greater than $\frac{1}{2}$ are taken as hindsight for long memory behavior. Therefore, the differencing parameter d can be estimated as $\hat{d} = \hat{H} - \frac{1}{2}$. The R/S-statistic can fairly easily be calculated in R, as shown in Rcode example 2.3.

Rcode 2.3 R/S statistic

```
library(fracdiff)                                        1
set.seed(123456)                                         2
# ARFIMA(0.0,0.3,0.0)                                     3
y <- fracdiff.sim(n=1000, ar=0.0, ma=0.0, d=0.3)         4
# Demean the series                                      5
y.dm <- y$series                                         6
max.y <- max(cumsum(y.dm))                               7
min.y <- min(cumsum(y.dm))                               8
sd.y <- sd(y$series)                                     9
RS <- (max.y - min.y)/sd.y                              10
H <- log(RS)/log(1000)                                  11
d <- H - 0.5                                            12
```

Because the default mean in the function `fracdiff` is zero, no demeaning has to be done. The estimated Hurst coefficient is 0.7843, which implies an estimated value for d of 0.2843 that is close to its simulated value of 0.3.

Since the seminal paper of Hurst, the rescaled range statistic has received intensive further research.[4] Although it has been long established that the R/S statistic has the ability to detect long-range dependence, it is however sensitive to short-range dependence and heteroskedasticity.[5] Hence, any incompatibility between the data and the predicted behavior of the R/S statistic under the null hypothesis of no long-run dependence need not come from long-term memory, but it may be merely a symptom of short-term autocorrelation. Lo [58] proposes a modified rescaled range statistic to cope with this deficiency. The modified R/S_{mod} is defined as

$$R/S_{mod} = \frac{1}{s_T(q)} \left[\max_{1 \le k \le T} \sum_{j=1}^{k} (y_j - \bar{y}) - \min_{1 \le k \le T} \sum_{j=1}^{k} (y_j - \bar{y}) \right] , \qquad (2.29)$$

where

$$s_T(q) = s_T + 2 \sum_{j=1}^{q} \omega_j(q) \hat{\gamma}_j , \, \omega_j(q) = 1 - \frac{j}{q+1} \text{ with } q < T . \qquad (2.30)$$

The maximum likehood standard deviation estimator is assigned by s_T and the jth-order sample autocorrelation by $\hat{\gamma}_j$. The sample autocorrelations are

---

[4] For instance, see Mandelbrot and Wallis [65][66] and Davies and Harte [15] who discuss alternative methods for estimating H. Anis and Loyd [2] determine the small sample bias.

[5] For instance, see Mandelbrot [63][64], Mandelbrot and Wallis [65], Davies and Harte [15], Aydogan and Booth [3], and Lo [58].

weighted by the function $w_j(q)$ proposed in Newey and West [70]. However, the choice of an appropriate order q is an unresolved issue.

A popular method for estimating d has been proposed by Geweke and Porter-Hudak [33]. They suggested a semiparametric estimator of d in the frequency domain. They consider as a data-generating process $(1-L)^d y_t = z_t$, where $z_t \sim I(0)$. This process can be represented in the frequency domain

$$f_y(\omega) = 1 - \exp(-i\omega)|^{-2d} f_z(\omega) , \tag{2.31}$$

where $f(\omega)_y$ and $f(\omega)_z$ assign the spectral densities of y_t and z_t, respectively. Equation 2.31 can be transformed to

$$\log\{f_y(\omega)\} = \{4\sin^2(\frac{\omega}{2})\}^{-d} + \log\{f_z(\omega)\} , \tag{2.32a}$$

$$\log\{f_y(\omega_j)\} = \log\{f_z(0)\} - d\log\{4\sin^2(\frac{\omega_j}{2})\} + \log\{\frac{f_u(\omega_j)}{f_z(0)}\} . \tag{2.32b}$$

The test regression is then a regression of the ordinates of the log spectral density on a trigonemetric function of frequencies:

$$\log\{I_y(\omega_j)\} = \beta_1 + \beta_2 \log\{4\sin^2(\frac{\omega_j}{2})\} + \nu_j , \tag{2.33}$$

where $\nu_j = \log\{\frac{f_z(\omega_j)}{f_z(0)}\}$ and $j = 1, \ldots, m$. The error term is assumed to be i.i.d. with zero mean and variance $\frac{\pi}{6}$. The estimated order of fractional differencing is equal to $\hat{d} = -\hat{\beta}_2$. Its significance can be tested with either the usual t ratio distributed as Student t, or one can set the residual variance equal to $\frac{\pi}{6}$. An example of this method is Rcode example 2.4, where a fractionally differenced series has been generated first with $d = 0.3$.

Rcode 2.4 Geweke and Porter-Hudak method

```
library(fracdiff)                                              1
set.seed(123456)                                               2
y <- fracdiff.sim(n=1000, ar=0.0, ma=0.0, d=0.3)               3
y.spec <- spectrum(y$series, plot=FALSE)                       4
lhs <- log(y.spec$spec)                                        5
rhs <- log(4*(sin(y.spec$freq/2))^2)                           6
gph.reg <- lm(lhs ~ rhs)                                        7
gph.sum <- summary(gph.reg)                                     8
sqrt(gph.sum$cov.unscaled*pi/6)[2,2]                            9
```

The results for the simulated fractionally differenced series are given in Table 2.1. The negative of the estimated coefficient $\hat{\beta}_2$ is 0.2968, which is close to its true value of $d = 0.3$ and higly significant on both accounts, i.e., its

	Estimate	Std. Error	t value	Pr(>|t|)
(Intercept)	−1.6173	0.1144	−14.14	0.0000
rhs	−0.2968	0.0294	−10.11	0.0000

Table 2.1. Results of Geweke and Porter-Hudak method

t ratio as well as the computed standard error with residual variance equal to $\frac{\pi}{6}$. Please note, a major issue with this approach is the selection of the range of frequencies to include in the regression. In Rcode example 2.4, all frequencies have been included, *i.e.*, 500. Diebold and Rudebusch [20] have set $m = \sqrt{T}$, Sowell [93] has suggested setting m to the shortest cycle associated with long-run behavior. A third possibility would be to choose m such that the estimated standard error of the regression is approximately equal to $\sqrt{\pi/6}$.

Finally, the estimation of an ARFIMA(p, d, q)–model is implemented in the contributed package `fracdiff` as function `fracdiff()`. The parameters are estimated by an approximated maximum likelihood using the method of Haslett and Raftery [39]. To lessen the computational burden, a range for the parameter d can be supplied as functional argument. In the case of a "less nonstationary" series, *i.e.*, $d > \frac{1}{2}$, the estimation fails and the series must be integer differenced first. In this case, the fractional differencing filter $(1 - L)^d$ is a combination of Equation 2.21 and integer differencing.

Summary

In this chapter, a more encompassing data-generating process that was introduced into the literature by Campbell and Perron [9] has been presented. You should now be familiar with the concepts of trend- *versus* difference-stationary and the decomposition of a time series into a deterministic trend, a stochastic trend, and a cyclical component. Furthermore, unit root processes have been introduced as a subclass of random walk processes. How one applies a sequential testing strategy to detect the underlying data-generating process of a possible nonstationary time series was discussed. The important definitions of integrated, seasonally integrated, and fractionally integrated time series processes have been presented too, whereas the latter can be viewed as a bridge between stationary and unit root processes, thereby closing the circle of the exposition in the first two chapters.

So far we have adressed univariate time series analysis only. The obstacles and solutions in a multivariate context are the subject of the next and last chapter of Part I.

Exercises

1. Write a function in R that returns the critical values given in Fuller [32]. As functional arguments should the test type, the level of significance and the sample size should be supplied.

2. Write a function in R that implements the ADF–test regression as shown in Equation 2.14. The series, the inclusion of a constant, trend, both or none, and the order of lagged differenced series should be included as functional arguments. The function should return a summary object of class lm.

3. Now include the function of Exercise 1 in the function of 2 such that the relevant critical values are returned beside a summary object of class lm.

4. Generate various long and intermediate processes for different values of d and AR(p) and MA(q) orders, and analyze their autocorrelation functions.

5. Write a function that estimates the Hurst coefficient, i.e., the R/S statistic as well as its modified version by Lo [58] and the order of the difference operator d.

6. Write a function for the single equation estimation of d as proposed by Geweke and Porter-Hudak [33].

7. Apply the functions of Exercises 5 and 6 to the absolute logarithmic returns of the stock indices contained in the data set EuStockMarkets. Can you detect long memory behavior in any of these series?

3

Cointegration

In the previous two chapters, a brief explanation of univariate time series models and their characteristics were presented. The focus of this chapter is on the simultaneous modeling of time series and inferences of the relationships between them. As will be shown, the degree of integration and a careful examination of the data-generating processes is of utmost importance. We will begin by briefly reviewing the case of a spurious regression before we proceed by providing a definition of cointegration and its error correction representation. In the last section, the more encompassing vector error correction models are presented.

3.1 Spurious Regression

Regression analysis plays a pivotal role in applied economics. It is widely used to test for the validity of economic theories. Furthermore, the *classic linear regression* models as in Equation 3.1 form the basis of macroeconomic forecasting and simulation models.

$$y_t = \beta_1 x_{t,1} + \beta_2 x_{t,2} + \ldots + \beta_K x_{t,K} + \varepsilon_t , \quad \text{for } t = 1, \ldots, T , \qquad (3.1)$$

where y_t assigns the endogenous variable, *i.e.*, the regressand; the exogenous variables, *i.e.*, the regressors, are included in the row vector $\boldsymbol{x}'_t = (x_{t,1}, x_{t,2}, \ldots, x_{t,K})$; and ε_t is a white noise random error. One important assumption of this model class is the stationarity of the variables, that is

$$\lim_{T \to \infty} \boldsymbol{X}'\boldsymbol{X} = \mathfrak{M} \text{ and } \exists\, \mathfrak{M}^{-1} . \qquad (3.2)$$

The product moment matrix of the regressors converges to fixed and invertible matrix $\mathfrak{M}$. This assumption is employed, *e.g.*, in the consistency proof of the ordinary least-squares (OLS) estimator. Clearly, for trending variables, as most often encountered in the field of empirical longitudal macroeconomic data, this assumption is not met. Incidentally, if only deterministic trends are present in the data-generating processes of the variables in question, then these can be removed before estimation of Equation 3.1 or can be included in the regression. The inference on the coefficients is the same regardless of which method is employed, *i.e.*, the Frisch–Waugh theorem [31]. However, matters are different in the case of difference-stationary data. In this case, the error term is often highly correlated and the t and F statistics are distorted such that the null hypothesis is rejected too often for a given critical value; hence, the risk of a "spurious regression" or "nonsense regression" exists.[1] Furthermore, such regressions are characterized by a high R^2. This fact arises because the endogenous variable contains a stochastic trend and the total variation is computed as $\sum_{t=1}^{T}(y_t - \bar{y})$; *i.e.*, it is erroneously assumed that the series has a fixed mean. Hence, given the formula for calculating the unadjusted R^2:

$$R^2 = 1 - \frac{\sum_{t=1}^{T} \hat{\varepsilon}_t^2}{\sum_{t=1}^{T}(y_t - \bar{y})^2} \, , \tag{3.3}$$

the goodness-of-fit measure tends to unity as the denominator becomes very large, because a large weight is placed to extreme observations on either side of the mean $\bar{y}$.

As a rule-of-thumb, Granger and Newbold [37] suggested that one should be suspicious if the R^2 is greater than the *Durbin–Watson statistic* [21][22][23]. A theoretical basis of their finding has been provided by Phillips [80].

In Rcode example 3.1, two unrelated random walk processes with drift have been generated and regressed on each other (see command line 8). The results are provided in Table 3.1.

Rcode 3.1 Spurious regression

```
library(lmtest)                          1
set.seed(123456)                         2
e1 <- rnorm(500)                         3
e2 <- rnorm(500)                         4
trd <- 1:500                             5
y1 <- 0.8*trd + cumsum(e1)               6
y2 <- 0.6*trd + cumsum(e2)               7
sr.reg <- lm(y1 ~ y2)                    8
sr.dw <- dwtest(sr.reg)$statistic        9
```

As can be seen, the coefficient of the regressor is significant, the adjusted R^2 of 0.9866 is close to one, and the Durbin–Watson statistic of 0.0172 is close to zero, as expected. For completeness, the Durbin–Watson statistic implemented in the contributed package lmtest has been used. An alternative is the durbin.watson() function in the contributed package car.

---

[1] The spurious regression problematic can be traced back to Yule [103] and Hooker [46]. For a historic background of nonsense regressions, see Hendry [43][42]. Hendry [41] has provided a pretty famous example of how easy it is to create a spurious regression by regressing the logarithm of the consumer price level on the cumulative rainfall in the United Kingdom.

	Estimate	Std. Error	t value	Pr(>|t|)
(Intercept)	−29.3270	1.3672	−21.45	0.0000
y2	1.4408	0.0075	191.62	0.0000

Table 3.1. Results of Spurious regression

From a statistical point of view, the spurious regression problem could be circumvented by taking first differences of the $I(1)$ variables in the regression equation and using these instead. However, by applying this procedure, two new problems are incurred. First, differencing greatly attenuates large positive residual autocorrelation; hence, false inferences upon the coefficients in the regression equation could be drawn. Second, most economic theories are expressed in levels, and the implications of the long-run relationships between variables are deducted. Therefore, being obliged to use regression approaches with differenced variables would be a great obstacle in the testing of economic theories. Other means of transforming nonstationary data into stationary ones, *e.g.*, by building logarithmic ratios, have been perused with success, for example, by Sargan [86] and Hendry and Anderson [44]. The reason why such a transformation is suitable in achieving stationarity is that the nonstationarities are "canceling" each other out, although this must not be true in all cases and all circumstances. All in all a new approach is called for dealing with trending variables in the context of regression analysis.

3.2 Concept of Cointegration and Error-Correction Models

In 1981, Granger [35] introduced the concept of *cointegration* into the literature, and the general case was publizised by Engle and Granger [26] in their seminal paper in 1987. The idea behind cointegration is to find a linear combination between two $I(d)$ variables that yields a variable with a lower order of integration. More formally, cointegration is defined as follows.

Definition 3.1. *The components of the vector $\boldsymbol{x}_t$ are said to be cointegrated of order d, b, denoted $\boldsymbol{x}_t \sim CI(d,b)$, if (a) all components of $\boldsymbol{x}_t$ are $I(d)$; and (b) a vector $\boldsymbol{\alpha}(\neq 0)$ exists so that $z_t = \boldsymbol{\alpha}'\boldsymbol{x}_t \sim I(d-b)$, $b > 0$. The vector $\boldsymbol{\alpha}$ is called the cointegrating vector.*

The great interest in this path-breaking development among economists is mostly explained by the fact that it is now possible to detect stable long-run relationships among nonstationary variables. Consider, the case of $d = 1$, $b = 1$; *i.e.*, the components in the vector $\boldsymbol{x}_t$ are all integrated of order one, but if a linear combination $\boldsymbol{\alpha}$ of these exists, then the resultant series z_t is stationary. Although the individual series are nonstationary, they are tied to each other by the cointegrating vector. In the parlance of economics, deviations from a long-run equilibrium path are possible, but these errors are characterized by a mean-reversion to its stable long-run equilibrium.

Now, the question is how to estimate the cointegrating vector α and how to model the dynamic behavior of $I(d)$ variables, in general, and for expositonal purposes of $I(1)$ variables, in particular?

Engle and Granger [26] proposed a *two-step* estimation technique to do so. In the first step, a regression of the variables in the set of $I(1)$ is run:

$$y_t = \alpha_1 x_{t,1} + \alpha_2 x_{t,2} + \ldots + \alpha_K x_{t,K} + z_t \text{ for } t = 1, \ldots, T, \qquad (3.4)$$

where z_t assigns the error term. The estimated $(K+1)$ cointegrating vector $\hat{\boldsymbol{\alpha}}$ is given by $\hat{\boldsymbol{\alpha}} = (1, -\hat{\boldsymbol{\alpha}}^*)'$, where $\hat{\boldsymbol{\alpha}}^* = (\hat{\alpha}_1, \ldots, \hat{\alpha}_K)'$. Hence, the cointegrating vector is normalized to the regressand. They showed that in this static regression, the cointegrating vector can be consistently estimated, but with a finite sample bias of magnitude $O_p(T^{-1})$. Because the usual convergence rate in the $I(0)$ case is only $O_p(T^{-1/2})$, Stock [95] termed the ordinary least-squares (OLS) estimation of the cointegrating vector as "super consistency." Incidentally, although the cointegrating vector can be super consistently estimated, Stock has shown that the limiting distribution is non-normal; hence, as in the case of spurious regressions, the typical t and F statistics are not applicable. However, what has been gained is first a resurrection of the applicability of the OLS method in the case of trending variables and second, the residuals from this static regression; *i.e.*, $\hat{z}_t$ are in the case of cointegration integrated of order zero. These residuals are the errors from the long-run equilibrium path of the set of $I(1)$ variables. Whether this series is stationary, *i.e.*, the variables are cointegrated, can be tested for example with the Dickey–Fuller (DF) test or the augmented Dickey–Fuller (ADF) test. Please note that now the critical values provided in Engle and Yoo [25] or in Phillips and Ouliaris [81] have to be considered because the series $\hat{z}_t$ is an estimated one.[2] As a rough check the so-called *cointegrating regression Durbin–Watson* test (henceforth: CRDW) proposed by Sargan and Bhargava [87] can be calculated with the null hypothesis CRDW $= 0$. The test statistic is the same as the usual Durbin–Watson test, but the prefix "cointegrating" has been added to emphasize its utilization in the context of cointegration testing. Once the null hypothesis of a unit root in the series $\hat{z}_t$ has been rejected, the second step of the two-step procedure follows. In this second step, an *error-correction-model* is specified (henceforth: ECM), *i.e.*, the Engle–Granger *representation theorem*. We restrict ourselves to the bivariate case first, in which two cointegrated variables y_t and x_t, each $I(1)$, are considered. In Section 3.3 systems of cointegrated variables are then presented. The general specification of an ECM is as follows:

---

[2] MacKinnon [62] has calculated critical values for the Dickey–Fuller (DF)– and augmented DF (ADF)–tests based on critical surface regressions.

$$\Delta y_t = \psi_0 + \gamma_1 \hat{z}_{t-1} + \sum_{i=1}^{K} \psi_{1,i} \Delta x_{t-i} + \sum_{i=1}^{L} \psi_{2,i} \Delta y_{t-i} + \varepsilon_{1,t} , \qquad (3.5a)$$

$$\Delta x_t = \xi_0 + \gamma_2 \hat{z}_{t-1} + \sum_{i=1}^{K} \xi_{1,i} \Delta y_{t-i} + \sum_{i=1}^{L} \xi_{2,i} \Delta x_{t-i} + \varepsilon_{2,t} , \qquad (3.5b)$$

where $\hat{z}_t$ is the error from the static regression in Equation 3.4, and with $\varepsilon_{1,t}$ and $\varepsilon_{2,t}$, white noise processes are assigned. The ECM as in equation 3.5a states that changes in y_t are explained by their own history, lagged changes of x_t, and the error from the long-run equilibrium in the previous period. The value of the coefficient γ_1 determines the speed of adjustment and should always be negative in sign. Otherwise the system would diverge from its long-run equilibrium path. Incidentally, one is not restricted by including the error from the previous period only. It can be any lagged value, because Equations 3.5a and 3.5b are still balanced because $\hat{z}_{t-1}$ is stationary and so is $\hat{z}_{t-k}$ with $k > 1$. Furthermore, as can be concluded from these equations and the static regression is that in the case of two cointegrated $I(1)$ variables, *Granger causality* must exist in at least one direction. That is, at least one variable can help forecast the other.

Aside from empirical examples of this method exhibited in Section 6.1, an artificial one is presented in the Rcode example 3.2.[3]

Rcode 3.2 Engle–Granger procedure with generated data

```
set.seed(123456)                                                      1
e1 <- rnorm(100)                                                      2
e2 <- rnorm(100)                                                      3
y1 <- cumsum(e1)                                                      4
y2 <- 0.6*y1 + e2                                                     5
lr.reg <- lm(y2 ~ y1)                                                 6
error <- residuals(lr.reg)                                            7
error.lagged <- error[-c(99, 100)]                                   8
dy1 <- diff(y1)                                                       9
dy2 <- diff(y2)                                                       10
diff.dat <- data.frame(embed(cbind(dy1, dy2), 2))                    11
colnames(diff.dat) <- c('dy1', 'dy2', 'dy1.1', 'dy2.1')              12
ecm.reg <- lm(dy2 ~ error.lagged + dy1.1 + dy2.1, data=diff.        13
    dat)
```

First two random walks have been created in which the latter one y2 has been set to 0.6*y2 + e2, where e2 is a white noise process (see command

---

[3] In the package's vignette of **strucchange**, an absolute consumption function for the United States is specified as an ECM.

lines 2 to 5). Hence, the cointegrating vector is $(1, -0.6)$. First, the long-run equation `lr.reg` has been estimated by OLS. Given a sample size of 100, as expected, the estimated coefficient of the regressor `y1` is close to its theoretical counterpart by having a value of 0.5811. In the next command lines, the equilibrium error is stored as `error` and its lagged version has been created by simply dropping the last two entries (see command line 8). Because differences and first lagged differences of `y1` and `y2` are generated with the commands `diff()` and `embed()`, subtracting the last two entries of the series `error` is equivalent with lagging the error term once in the ensuing ECM-regression (see command line 13). The results of the ECM are displayed in

	Estimate	Std. Error	t value	Pr($>$|t|)
(Intercept)	0.0034	0.1036	0.03	0.9739
error.lagged	−0.9688	0.1586	−6.11	0.0000
dy1.1	0.8086	0.1120	7.22	0.0000
dy2.1	−1.0589	0.1084	−9.77	0.0000

Table 3.2. Results of Engle–Granger procedure with generated data

Table 3.2. To no suprise, the equilibrium error of the last period is almost completely worked off. Its coefficient is significant and close to negative one. So far, we have restricted the exposition to the bivariate case and hence to only one cointegrating vector. However, if the dimension n of the vector x'_t is greater than two, at most $n - 1$ distinct linear combinations could exist that would each produce a series with a lower order of integration. Therefore, by applying the above-presented Engle–Granger two-step procedure in a case in which $n > 2$, one estimates a single cointegrating vector only, which would represent an average of up to $n - 1$ distinct cointegrating relationships. How to cope with multiple long-run relationships is the subject of Section 3.3.

3.3 Systems of Cointegrated Variables

Before the *vector error correction model* (henceforth: VECM) is presented, the time series decomposition in a deterministic and a stochastic component as in Equation 2.1 is extended to the multivariate case and the broader concept of cointegration is defined.

We now assume that each component of the $(n \times 1)$ vector y_t, for $t = 1, \ldots, T$, can be represented as

$$y_{i,t} = TD_{i,t} + z_{i,t} \text{ for } i = 1, \ldots, n \text{ and } t = 1, \ldots, T , \qquad (3.6)$$

where $TD_{i,t}$ assigns the deterministic component of the ith variable and $z_{i,t}$ represents the stochastic component as an autoregressive-moving average (ARMA)–process, $\phi_i(L)z_{i,t} = \theta_i(L)\varepsilon_{i,t}$. It is further assumed that $y_{i,t}$

contains maximal one unit root and all remaining ones are outside the unit circle.

Campbell and Perron [9] have then defined cointegration in a broader sense as follows.

Definition 3.2. *An $(n \times 1)$ vector of variables $\boldsymbol{y}_t$ is said to be cointegrated if at least one nonzero n-element vector $\boldsymbol{\beta}_i$ exists such that $\boldsymbol{\beta}'_i \boldsymbol{y}_t$ is trend station-ary. $\boldsymbol{\beta}_i$ is called a cointegrating vector. If r such linearly independent vectors $\boldsymbol{\beta}_i (i = 1, \ldots, r)$ exist, we say that $\{\boldsymbol{y}_t\}$ is cointegrated with cointegrating rank r. We then define the $(n \times r)$-matrix of cointegrating vectors $\boldsymbol{\beta} = (\boldsymbol{\beta}_1, \ldots, \boldsymbol{\beta}_r)$. The r elements of the vector $\boldsymbol{\beta}' \boldsymbol{y}_t$ are trend-stationary and $\boldsymbol{\beta}$ is called the coin-tegrating matrix.*

This definition is broader than the one by Engle and Granger (see Defini-tion 3.1) in the sense that now it is not any longer required that each individual series be integrated of the same order. For example, some or all series can be trend stationary. If $\boldsymbol{y}_t$ contains a trend-stationary variable, then it is triv-ially cointegrated and the cointegrating vector is the unit vector that selects the stationary variable. On the other hand, if all series are trend-stationary, then the system is again trivially cointegrated because any linear combination of trend-stationary variables yields a trend-stationary variable. Furthermore, nonzero linear trends in the data are also included per Equation 3.6.

Consider now, the following vector autoregression model of order K (hence-forth: VAR):

$$\boldsymbol{y}_t = \boldsymbol{\Pi}_1 \boldsymbol{y}_{t-1} + \ldots + \boldsymbol{\Pi}_K \boldsymbol{y}_{t-K} + \boldsymbol{\mu} + \boldsymbol{\Phi} \boldsymbol{D}_t + \boldsymbol{\varepsilon}_t \text{ for } t = 1, \ldots, T), , \quad (3.7)$$

where $\boldsymbol{y}_t$ assigns the $(n \times 1)$ vector of series at period t, the matrices $\boldsymbol{\Pi}_i (i = 1, \ldots, K)$ are the $(n \times n)$ coefficient matrices of the lagged endogenous variables, $\boldsymbol{\mu}$ is a $(n \times 1)$ vector of constants, and $\boldsymbol{D}_t$ is a vector of nonstochas-tic variables, such as seasonal dummies or intervention dummies. The $(n \times 1)$ error term $\boldsymbol{\varepsilon}_t$ is assumed to be i.i.d. as $\boldsymbol{\varepsilon}_t \sim \mathcal{N}(\boldsymbol{0}, \boldsymbol{\Sigma})$.

As long as all components of $\boldsymbol{y}_t$ are (trend)-stationary, the above equations can be estimated separately or as a system by maximum-likelihood or OLS. The estimation of VAR-models is implemented in the contributed package dse1 as functions estVARXls() and estVARXar().

From Equation 3.7, two versions of a VECM can be delineated. In the first form, the levels of $\boldsymbol{y}_t$ enter with lag $t - K$:

$$\Delta \boldsymbol{y}_t = \boldsymbol{\Gamma}_1 \Delta \boldsymbol{y}_{t-1} + \ldots + \boldsymbol{\Gamma}_{K-1} \Delta \boldsymbol{y}_{t-K+1} + \boldsymbol{\Pi} \boldsymbol{y}_{t-K} + \boldsymbol{\mu} + \boldsymbol{\Phi} \boldsymbol{D}_t + \boldsymbol{\varepsilon}_t , \quad (3.8a)$$

$$\boldsymbol{\Gamma}_i = -(\boldsymbol{I} - \boldsymbol{\Pi}_1 - \ldots - \boldsymbol{\Pi}_i) , \text{ for } i = 1, \ldots, K - 1 , \quad (3.8b)$$

$$\boldsymbol{\Pi} = -(\boldsymbol{I} - \boldsymbol{\Pi}_1 - \cdots - \boldsymbol{\Pi}_K) , \quad (3.8c)$$

where I is the $(n \times n)$ identity matrix. As can be seen from Equation 3.8b, the $\Gamma_i (i = 1, \ldots, K-1)$ matrices contain the cumulative long-run impacts; hence, this specification is termed the *long-run form*. Please note that the levels of y_t enter with lag $t - K$.

The other VECM specification is of the form:

$$\Delta y_t = \Gamma_1 \Delta y_{t-1} + \ldots + \Gamma_{K-1} \Delta y_{t-K+1} + \Pi y_{t-1} + \mu + \Phi D_t + \varepsilon_t , \quad (3.9a)$$
$$\Gamma_i = -(\Pi_{i+1} + \ldots + \Pi_K) , \text{ for } i = 1, \ldots, K-1 , \quad (3.9b)$$
$$\Pi = -(I - \Pi_1 - \cdots - \Pi_K) . \quad (3.9c)$$

The Π matrix is the same as in the first specification. However, the Γ_i matrices now differ, in the sense that they measure transitory effects; hence, this form of the VECM is termed the *transitory form*. Furthermore, the levels of the components in y_t enter lagged by one period. Incidentally, as will become evident, inferences drawn on Π will be the same, regardless of which specification is chosen and the explanatory power is the same.

Per assumption, the individual components of y_t are at most $I(1)$ variables (see Definition 3.2). Therefore, the left-hand side of the VECM is stationary. Beside lagged differences of y_t, the error-correction term Πy_{t-K}, or, dependent on the specification of the VECM, Πy_{t-1} appears. This term must be stationary too; otherwise the VECM would not balance. The question now is, what kind of conditions must be given for the matrix Π such that the right-hand is stationary? Three cases must be considered:

i) $rk(\Pi) = n,$
ii) $rk(\Pi) = 0 ,$
iii) $0 < rk(\Pi) = r < n,$

where $rk()$ assigns the rank of a matrix. In the first case, all n linearly independent combinations must be stationary. It can only be the case if the deviations of y_t around the deterministic components are stationary. Equations 3.8 and 3.9 represent a standard VAR-model in levels of y_t. In the second case, in which the rank of Π is zero, no linear combination exists to make Πy_t stationary, except for the trivial solution. Hence, this case would correspond to a VAR-model in first differences. The interesting case is the third one in which $0 < rk(\Pi) = r < n$. Because the matrix does not have full rank, two $(n \times r)$ matrices α and β exist such that $\Pi = \alpha\beta'$. Hence, $\alpha\beta'y_{t-K}$ is stationary and therefore the matrix-vector product $\beta'y_{t-K}$ is stationary. The r linear independent columns of β are the cointegrating vectors and the rank of Π is equal to the cointegration rank of the system y_t. That is, each column represents one long-run relationship between the individual series of y_t. However, the parameters of the matrices α and β are undefined because any nonsingular matrix Ξ would yield $\alpha\Xi(\beta\Xi^{-1})' = \Pi$. It implies that only the cointegration space spanned by $\boldsymbol{\beta}$ can be determined. The obvious solution is to normalize one element of $\boldsymbol{\beta}$ to one. The elements of $\boldsymbol{\alpha}$ determine the speed of adjustment to the long-run equilibrium. It is referred to as the *loading* or *adjustment* matrix.

Johansen [51][52] and Johansen and Juselius [54] developed maximum likelihood estimators of these cointegration vectors for an autoregressive process as in Equations 3.7 through 3.9. A thorough and concise presentation of this approach is given in a monograph by Johansen [53]. Their approach uses *canonical correlation* analysis as a means to reduce the information content of T observations in the n dimensional space to a lower dimensional one of r cointegrating vectors. Hence, the canonical correlations determine to what extent the multicollinearity in the data will allow such a smaller r dimensional space. To do so, $2n$ auxilliary regressions are estimated by OLS: $\Delta \boldsymbol{y}_t$ is regressed on lagged differences of $\boldsymbol{y}_t$. The residuals are termed $\boldsymbol{R}_{0t}$. In the second set of auxilliary regressions, $\boldsymbol{y}_{t-K}$ is regressed on the same set of regressors. Here, the residuals are assigned as $\boldsymbol{R}_{kt}$. The $2n$ residuals series of these regression are used to compute the product moment matrices as

$$\hat{\boldsymbol{S}}_{ij} = \frac{1}{T} \sum_{t=1}^{T} \boldsymbol{R}_{it} \boldsymbol{R}'_{jt} \text{ with } i, j = 0, k . \tag{3.10}$$

Johansen showed that the likelihood ratio test statistic of the null hypothesis that there are at most r cointegrating vectors is

$$-2\ln(Q) = -T \sum_{i=r+1}^{n} (1 - \hat{\lambda}_i) , \tag{3.11}$$

where $\hat{\lambda}_{r+1}, \dots, \hat{\lambda}_p$ are the $n - r$ smallest eigenvalues of the equation:

$$\left| \lambda \hat{\boldsymbol{S}}_{kk} - \hat{\boldsymbol{S}}_{k0} \hat{\boldsymbol{S}}_{00}^{-1} \hat{\boldsymbol{S}}_{0k} \right| = 0 . \tag{3.12}$$

For ease of computation, the $(n \times n)$ matrix $\hat{\boldsymbol{S}}_{kk}$ can be decomposed into the product of a nonsingular $(n \times n)$ matrix $\boldsymbol{C}$, such that $\hat{\boldsymbol{S}}_{kk} = \boldsymbol{C}\boldsymbol{C}'$. Equation 3.12 would then accordingly be written as

$$\left| \lambda \boldsymbol{I} - \boldsymbol{C}^{-1} \hat{\boldsymbol{S}}_{k0} \hat{\boldsymbol{S}}_{00}^{-1} \hat{\boldsymbol{S}}_{0k} \boldsymbol{C}'^{-1} \right| = 0 , \tag{3.13}$$

where $\boldsymbol{I}$ assigns the identity matrix.
Johansen [51] has tabulated critical values for the test statistic in Equation 3.11 for various quantiles and up to five cointegration relations, *i.e.*, $r = 1, \dots, 5$. This statistic has been named the *trace statistic*.
Beside the trace statistic, Johansen [54] has suggested the *maximal eigenvalue statistic* defined as

$$-2\ln(Q; r|r+1) = -T \ln(1 - \hat{\lambda}_{r+1}) , \tag{3.14}$$

for testing the existence of r versus $r+1$ cointegration relationships. Critical values for both test statistics and different specifications with respect to the inclusion of deterministic regressors are provided in the appendix of Johansen and Juselius [54].

Once the cointegration rank r has been determined, the cointegrating vectors can be estimated as

$$\hat{\boldsymbol{\beta}} = (\hat{\boldsymbol{v}}_1, \ldots, \hat{\boldsymbol{v}}_r) \,, \tag{3.15}$$

where $\hat{\boldsymbol{v}}_i$ are given by $\hat{\boldsymbol{v}}_i = \boldsymbol{C}'^{-1}\boldsymbol{e}_i$ and $\boldsymbol{e}_i$ are the eigenvectors to the corresponding eigenvalues in Equation 3.13. Equivalent to this are the first r eigenvectors of $\hat{\boldsymbol{\lambda}}$ in Equation 3.12 if they are normalized such that $\hat{\boldsymbol{V}}'\hat{\boldsymbol{S}}_{kk}\hat{\boldsymbol{V}} = \boldsymbol{I}$ with $\hat{\boldsymbol{V}} = (\hat{\boldsymbol{v}}_1, \ldots, \hat{\boldsymbol{v}}_n)$.

The adjustment matrix $\boldsymbol{\alpha}$ is estimated as

$$\hat{\boldsymbol{\alpha}} = -\hat{\boldsymbol{S}}_{0k}\hat{\boldsymbol{\beta}}(\hat{\boldsymbol{\beta}}'\hat{\boldsymbol{S}}_{kk}\hat{\boldsymbol{\beta}})^{-1} = -\hat{\boldsymbol{S}}_{0k}\hat{\boldsymbol{\beta}} \,. \tag{3.16}$$

The estimator for $\boldsymbol{\alpha}$ is dependent on the choice of the optimizing $\boldsymbol{\beta}$. The estimator for the matrix $\boldsymbol{\Pi}$ is given as

$$\hat{\boldsymbol{\Pi}} = -\hat{\boldsymbol{S}}_{0k}\hat{\boldsymbol{\beta}}(\hat{\boldsymbol{\beta}}'\hat{\boldsymbol{S}}_{kk}\hat{\boldsymbol{\beta}})^{-1}\hat{\boldsymbol{\beta}}' = -\hat{\boldsymbol{S}}_{0k}\hat{\boldsymbol{\beta}}\hat{\boldsymbol{\beta}}' \,. \tag{3.17}$$

Finally, the variance-covariance matrix of the n dimensional error process $\boldsymbol{\varepsilon}_t$ is given as

$$\hat{\boldsymbol{\Sigma}} = \hat{\boldsymbol{S}}_{00} - \hat{\boldsymbol{S}}_{0k}\hat{\boldsymbol{\beta}}\hat{\boldsymbol{\beta}}'\hat{\boldsymbol{S}}_{k0} = \hat{\boldsymbol{S}}_{00} - \hat{\boldsymbol{\alpha}}\hat{\boldsymbol{\alpha}}' \,. \tag{3.18}$$

The first part ends with a three-dimensional example of the above briefly exhibited Johansen procedure. The estimation and testing of the cointegration rank in a VECM is implemented in the package urca as ca.jo().[4] Besides these two functionalities, the testing of restrictions based on $\boldsymbol{\alpha}$ or $\boldsymbol{\beta}$ or on both as well as the validity of deterministic regressors will presented in Section 7.1.

In Rcode example 3.3 two random walks, x and y, have been generated with a size of 100 and combined by the cointegration vector $(1, 0.4)$. The VECM has been estimated with the function ca.jo()(see command line 8). The default value for the test statistic is the maximum eigenvalue test. The results are provided in Table 3.3.

---

[4] Incidentally, a graphical user interface for the package urca is shipped in the inst subdirectory of the package as an add-in to the graphical user interface Rcmdr by Fox [28]. It is particularly suited for teaching purposes, as the apprentice can concentrate on the methodology and is at the beginning not distracted by the function's syntax.

Rcode 3.3 Johansen method with artificially generated data

```
library(urca)                                          1
set.seed(123456)                                       2
e1 <- rnorm(100)                                       3
e2 <- rnorm(100)                                        4
x <- cumsum(e1)                                         5
y <- -0.4*x + e2                                        6
simdat <- cbind(y, x)                                   7
jo.results <- summary(ca.jo(simdat))                   8
class(jo.results)                                       9
slotNames(jo.results)                                  10
```

	test statistic	10%	5&	1%
r <= 1	2.05	2.82	3.96	6.94
r = 0	36.43	12.10	14.04	17.94

Table 3.3. Cointegration rank: Maximum eigenvalue statistic

Clearly, the hypothesis of one cointegrating vector has to be rejected for all significance levels. Table 3.4, the estimated cointegrating vectors are displayed. In the first column, the cointegration vector is the one associated with the largest eigenvalue. The value is close to the theoretical one. The second column can be disregarded because, first, we are working with a bivariate model and hence only one cointegration vector can exist and, second, even if we would have included more variables, the only interesting cointegration vectors are the first r ones (reading from left to right), where the cointegration rank has been determined by either the trace or the maximum eigenvalue test.

	1	2
y	1.00	1.00
x	0.42	12.23

Table 3.4. Cointegration vectors

This chapter ends with some R technicalities about the package **urca**. In this package, $S4$-classes are employed, in contrast to the older $S3$-classes. The main difference for the user is that "slots" of an object that belong to a certain class cannot be retrieved with `object$slot` as usual, but one has to use the @ sign instead. Furthermore, the slots of an $S4$-class object cannot be shown with `names(object)` as is the case with *S3*-class objects, but as shown in Rcode example 3.3 with the command `slotNames()`. The name of the class an object belongs to can be retrieved by `class(object)`. Although being very brief on this topic, the reader is referred to the documentation of the `methods` package [83] as well as to Chambers [11], but for applying the functions contained in `urca`, this background information should be sufficient.

Summary

By outlining the spurious regression problem in the first section you should have been alerted to the pitfalls, when integrated time series are modeled in a multivariate context; recall the rule-of-thumb for detecting such nonsense regressions. In the ensuing section, the imminent solution to this problem was presented, namely the definition of cointegration as a a linear combination with a degree of lower integratedness than the two integrated processes to be investigated. In this respect, it has been pointed out first, that if two time series are cointegrated, then an error-correction mechanism exists and *vice versa* and second, that in the case of two cointegrated $I(1)$ variables, Granger causality must exist in at least one direction. An empirical example of these issues is presented in Section 6.1. The short-coming of the Engle–Granger procedure is that in the case of more than two integrated variables not only one cointegrating vector can exist. In fact, by applying the Engle–Granger two-step procedure in cases with more than one cointegrating vector, one would estimate a linear combination of these vectors. An encompassing defintion of cointegration and the model class of VECMS has been introduced to cope adequately with such instances. It has been shown that two forms of a VECM exist and that the inference with regard to the order of the space spanned by the linearly independent cointegrating vectors is the same. You should recall that neither the cointegrating vectors nor the adjustment matrix can uniquely be determined. Instead a normalization of one element of β to one is applied.

By now, the likelihood-based inference in cointegrated vector autoregressive models has been confined to determining the cointegration rank only. Testing various restrictions placed on the cointegration vectors and the loading matrix as well as a combination thereof are presented in Sections 7.1.3 and 7.1.4.

Exercises

1. Write a function in R that returns the critical values for the cointegration unit root tests as given in Engle and Yoo [25], Phillips and Ouliaris [72], and MacKinnon [62]. As functional arguments, the relevant tables, the sample size, and where applicable the number of variables in the long-run relationships should be supplied.

2. Write a function in R that implements the Engle–Granger two-step method as shown in Equations 3.4 and 3.5. The series, and the order of lagged differenced series, should be included as functional arguments. The function should return a summary object of class `lm`.

3. Now include the function of Exercise 1 in the function of Exercise 2 such that the relevant critical value is returned beside a summary object of class `lm`.

Part II

Unit Root Tests

Testing for the Order of Integration

This chapter is the first in which the theoretical aspects laid out in Part I of the book are put into "practice." We begin by introducing the most commomly employed unit root tests in econometrics: the Dickey–Fuller test and its extensions. To discriminate between trend and difference stationary time series processes, a sequential testing strategy is described. Other unit root tests encountered in applied research are presented in the ensuing sections.

4.1 Dickey–Fuller–Type Tests

We now apply the augmented Dickey–Fuller (ADF) test to the data sample used by Holden and Perman [45]. The authors applied an integration/cointegration analysis to a consumption function for the United Kingdom using quarterly data for the period 1966:Q4-1991:Q2. This data set is included in the contributed package urca as Raotbl3. The consumption series is seasonally adjusted real consumer expenditure in prices of 1985. The seasonal adjusted personal disposable income series has been deflated by the implicit consumption price index; likewise the wealth series that is defined as seasonally adjusted gross personal financial wealth. All variables are expressed in their natural logarithms. Recall from Section 2.2 the test regression 2.15, which is reprinted here with the three different combinations of the deterministic part:

$$\Delta y_t = \beta_1 + \beta_2 t + \pi y_{t-1} + \sum_{j=1}^{k} \gamma_j \Delta y_{t-j} + u_{1t} , \tag{4.1a}$$

$$\Delta y_t = \beta_1 + \pi y_{t-1} + \sum_{j=1}^{k} \gamma_j \Delta y_{t-j} + u_{2t} , \tag{4.1b}$$

$$\Delta y_t = \pi y_{t-1} + \sum_{j=1}^{k} \gamma_j \Delta y_{t-j} + u_{3t} . \tag{4.1c}$$

The ADF–test has been implemented in the contributed packages fSeries, tseries, urca, and uroot as functions adftest(), adf.test(), ur.df(), and ADF.test(), respectively. For determing the integration order as outlined in Section 2.2, we will use the function ur.df() for the consumption series. The reason for this is twofold. First, the three different specifications as in Equations 4.1a–4.1c can be modeled which is not the case for the function adf.test(), and second, beside the τ statistics, the F type statistics are returned in the slot object@teststat with their critical values in the slot object@cval, as we will see shortly.

In Rcode example 4.1, the test regressions for Models 4.1a and 4.1b are estimated.

Rcode 4.1 ADF–test: Integration order for consumption in the United Kingdom

```
library(urca)                                                      1
data(Raotbl3)                                                      2
attach(Raotbl3)                                                    3
lc <- ts(lc, start=c(1966,4), end=c(1991,2), frequency=4)         4
lc.ct <- ur.df(lc, lags=3, type='trend')                          5
plot(lc.ct)                                                        6
lc.co <- ur.df(lc, lags=3, type='drift')                          7
lc2 <- diff(lc)                                                    8
lc2.ct <- ur.df(lc2, type='trend', lags=3)                        9
```

As a first step a regression with a constant and a trend has been estimated (see command line 5). Three lagged endogenous variables have been included to assure a spherical error process as is witnessed by the autocorrelations and partial autocorrelations in Figure 4.1. Including a fourth lag turns out to be insignificant, whereas specifiying the test regression with only two lagged endogenous variables does not suffice to achieve serially uncorrelated errors. The summary output of this test regression is provided in Table 4.1. Next, the hypothesis $\phi_3 = (\beta_1, \beta_2, \pi) = (\beta_1, 0, 1)$ is tested by a usual F type test. That is, zero restrictions are placed on the time trend and the lagged value of lc. The result is displayed in Table 4.2. The test statistic has a value of 2.6. Please

	Estimate	Std. Error	t value	Pr(>|t|)
(Intercept)	0.7977	0.3548	2.25	0.0270
z.lag.1	−0.0759	0.0339	−2.24	0.0277
tt	0.0005	0.0002	2.28	0.0252
z.diff.lag1	−0.1064	0.1007	−1.06	0.2934
z.diff.lag2	0.2011	0.1012	1.99	0.0500
z.diff.lag3	0.2999	0.1021	2.94	0.0042

Table 4.1. ADF–test: Regression for consumption with constant and trend

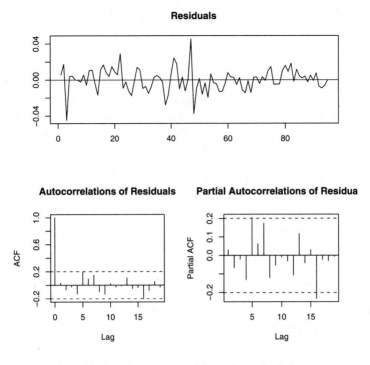

Fig. 4.1. ADF–test: Diagram of fit and residual diagnostics

note that one must consult the critical values in Dickey and Fuller [17], Table VI. The critical value for a sample size of 100 and significance levels of 10%, 5%, and 1% are 5.47, 6.49, and 8.73, respectively. Hence, the null hypothesis cannot be rejected, which implies that real consumption does contain a unit root. This finding is reiterated by a t ratio of –2.24 for the lagged endogenous

	test statistic	1%	5%	10%
τ_3	−2.24	−4.04	−3.45	−3.15
ϕ_2	3.74	6.50	4.88	4.16
ϕ_3	2.60	8.73	6.49	5.47

Table 4.2. ADF–test: τ_3, ϕ_2, and ϕ_3 test

variable in levels. The relevant critical values have now to be taken from Fuller [32], Table 8.5.2, which is given for a sample size of 100 and a signifance level of 5% equal to -3.45.[1]

---

[1] Instead of using the criticial values in Fuller [32], one can employ the same ones provided in Hamilton [38] or the ones calculated by critical surface regressions in

Therefore, a unit root cannot be rejected. Next, it is tested whether the consumption series is a random walk with or without drift (see command line 7). The relevant test statistic is ϕ_2, which is provided in Table 4.2. The value of this test is 3.74 and has to be compared with the critical values provided in Dickey and Fuller [17], Table V. For a sample size of 100, these values are 4.16, 4.88, and 6.50 for significance levels of 10%, 5%, and 1%, respectively. The conclusion is that the consumption series behaves like a pure random walk. One proceeds next by estimating Equation 4.1b based on the result of the ϕ_3 test. The results are depicted in Table 4.3. For the sake of completeness, it is now tested whether in this model a drift term is absent. The test results are provided in Table 4.4. The test statistic is 2.88, which turns out to be insignificant compared with the critical values of Table IV in Dickey and Fuller [17]. Therefore, the conclusion is that the quarterly real consumption

	Estimate	Std. Error	t value	Pr(>|t|)
(Intercept)	0.0123	0.0851	0.14	0.8852
z.lag.1	−0.0007	0.0079	−0.09	0.9261
z.diff.lag1	−0.1433	0.1016	−1.41	0.1620
z.diff.lag2	0.1615	0.1020	1.58	0.1169
z.diff.lag3	0.2585	0.1027	2.52	0.0136

Table 4.3. ADF–test: Regression for consumption with constant only

	test statistic	1%	5%	10%
τ_2	−0.09	−3.51	−2.89	−2.58
ϕ_1	2.88	6.70	4.71	3.86

Table 4.4. ADF–test: τ_2 and ϕ_1 test

series does contain a unit root, but neither a linear trend nor a drift is present in the data-generating process.

Finally, it is tested whether differencing the series once suffices to achieve stationarity; *i.e.*, it is tested whether the series is possible $I(2)$. This test is achieved by supplying the differenced series as argument y in the function ur.df as is done in the last two command lines of Rcode example 4.1. The results are displayed in Table 4.5. The hypothesis that the consumption series is $I(2)$ must be clearly dismissed given a t ratio of –4.39.

MacKinnon [62]. In the function ur.df(), the critical values provided in Hamilton [38] have been implemented for the τ statistics and the ones provided in Dickey and Fuller [17] for the ϕ statistics. These are stored in the slot object@cval.

	Estimate	Std. Error	t value	Pr(>|t|)
(Intercept)	0.0039	0.0031	1.27	0.2087
z.lag.1	−0.8826	0.2013	−4.39	0.0000
tt	0.0000	0.0001	0.62	0.5348
z.diff.lag1	−0.2253	0.1873	−1.20	0.2321
z.diff.lag2	−0.0467	0.1600	−0.29	0.7711
z.diff.lag3	0.1775	0.1057	1.68	0.0967

Table 4.5. ADF–test: Regression for testing $I(2)$

4.2 Phillips–Perron Test

Phillips and Perron [82] and Perron [74] suggested a nonparametric test statistics for the null hypothesis of a unit root that explicitly allow for weak dependence and heterogenity of the error process (henceforth: PP–test). The authors consider the following two test regressions:

$$y_t = \mu + \alpha y_{t-1} + \varepsilon_t , \qquad (4.2a)$$

$$y_t = \mu + \beta(t - \frac{1}{2}T) + \alpha y_{t-1} + \varepsilon_t . \qquad (4.2b)$$

They define the following test statistics for Equation 4.2a:

$$Z(\hat{\alpha}) = T(\hat{\alpha} - 1) - \hat{\lambda}/\bar{m}_{yy} , \qquad (4.3a)$$

$$Z(\tau_{\hat{\alpha}}) = (\hat{s}/\hat{\sigma}_{Tl})t_{\hat{\alpha}} - \hat{\lambda}'\hat{\sigma}_{Tl}/\bar{m}_{yy}^{\frac{1}{2}} , \qquad (4.3b)$$

$$Z(\tau_{\hat{\mu}}) = (\hat{s}/\hat{\sigma}_{Tl})t_{\hat{\mu}} + \hat{\lambda}'\hat{\sigma}_{Tl}m_y/\bar{m}_{yy}^{\frac{1}{2}}m_{yy}^{\frac{1}{2}} , \qquad (4.3c)$$

with $\bar{m}_{yy} = T^{-2}\sum(y_t - \bar{y})^2$, $m_{yy} = T^{-2}\sum y_t^2$, $m_y = T^{-3/2}\sum y_t$, and $\hat{\lambda} = 0.5(\hat{\sigma}_{Tl}^2 - \hat{s}^2)$, where $\hat{s}^2$ is the sample variance of the residuals, $\hat{\lambda}' = \hat{\lambda}/\hat{\sigma}_{Tl}^2$, and $t_{\hat{\alpha}}$, $t_{\hat{\mu}}$ are the t ratios of $\hat{\alpha}$ and $\hat{\mu}$, respectively. The long-run variance $\hat{\sigma}_{Tl}^2$ is estimated as

$$\hat{\sigma}_{Tl}^2 = T^{-1}\sum_{t=1}^{T}\hat{\varepsilon}_t^2 + 2T^{-1}\sum_{s=1}^{l}w_{sl}\sum_{t=s+1}^{T}\hat{\varepsilon}_t\hat{\varepsilon}_{t-s} , \qquad (4.4)$$

where $w_{sl} = 1 - s/(l+1)$.

Similarily, the following test statistics are defined for the test regression with a linear time trend included as in Equation 4.2b:

$$Z(\tilde{\alpha}) = T(\tilde{\alpha} - 1) - \tilde{\lambda}/M , \qquad (4.5a)$$

$$Z(t_{\tilde{\alpha}}) = (\tilde{s}/\tilde{\sigma}_{Tl})t_{\tilde{\alpha}} - \tilde{\lambda}'\tilde{\sigma}_{Tl}/M^{\frac{1}{2}} , \qquad (4.5b)$$

$$Z(t_{\tilde{\mu}}) = (\tilde{s}/\tilde{\sigma}_{Tl})t_{\tilde{\mu}} - \tilde{\lambda}'\tilde{\sigma}_{Tl}m_y/M^{\frac{1}{2}}(M + m_y^2)^{\frac{1}{2}} , \qquad (4.5c)$$

$$Z(t_{\tilde{\beta}}) = (\tilde{s}/\tilde{\sigma}_{Tl})t_{\tilde{\beta}} - \tilde{\lambda}'\tilde{\sigma}_{Tl}(\frac{1}{2}m_y - m_{ty})/(M/12)^{\frac{1}{2}}\bar{m}_{yy}^{\frac{1}{2}} , \qquad (4.5d)$$

where m_y, $\bar{m}_{yy}$, $\tilde{\lambda}$, $\tilde{\lambda}'$, and $\tilde{\sigma}_{Tl}$ are defined likewise as in Equations 4.3a-4.3c, $m_{ty} = T^{-5/2} \sum t y_t$, $t_{\tilde{\mu}}$, $t_{\tilde{\beta}}$, and $t_{\tilde{\alpha}}$ are the t ratios of $\tilde{\mu}$, $\tilde{\alpha}$, and $\tilde{\beta}$, respectively. The scalar M is defined as $M = (1 - T^{-2})m_{yy} - 12m_{ty}^2 + 12(1 + T^{-1})m_{ty}m_y - (4 + 6T^{-1} + 2T^{-2})m_y^2$.

The critical values of these Z statistics are identical to the ones of the DF–type tests. The advantage is that these modified tests eliminate the nuisance parameters that are present in the DF statistic if the error process does not satisfy the i.i.d assumption. However, one problem with these tests is that it is at the researcher's discretion to choose an optimal lag number l for computing the long-run variances $\hat{\sigma}_{Tl}^2$ or $\tilde{\sigma}_{Tl}^2$ as in Equation 4.4.

The PP–test is implemented as function `pp.test()` in the contributed package `tseries` and as function `ur.pp()` in the contributed package `urca`.[2] The advantage of the latter is that all test statistics are computed and returned by applying the summary method to an object of class `ur.pp` as well as to the test regression. Furthermore, the lags to be included in the computation of the long-run variance can be set either manually via the argument `use.lag` or it can be chosen automatically via the argument `lags` to be `short` or `long`, which corresponds to the integer values of $4(T/100)^{\frac{1}{4}}$ and $12(T/100)^{\frac{1}{4}}$, respectively.

In Rcode example 4.2, the PP–test is applied to the consumption series used in Rcode example 4.1.

Rcode 4.2 PP–test: Integration order for consumption in the United Kingdom

```
library(urca)                                                          1
data(Raotbl3)                                                          2
attach(Raotbl3)                                                        3
lc <- ts(lc, start=c(1966,4), end=c(1991,2), frequency=4)             4
lc.ct <- ur.pp(lc, type='Z-tau', model='trend', lags='long')         5
lc.co <- ur.pp(lc, type='Z-tau', model='constant', lags='          6
    long')
lc2 <- diff(lc)                                                        7
lc2.ct <- ur.pp(lc2, type='Z-tau', model='trend', lags='long       8
    ')
```

---

[2] Both functions have also been ported into the package `fSeries` as functions `tsppTest()` and `urppTest()`, respectively.

First the test is applied to Equation 4.2b, and the results are stored in the object lc.ct (see command line 5). The result of the test regression is displayed in Table 4.6.

	Estimate	Std. Error	t value	Pr(>|t|)
(Intercept)	0.5792	0.3622	1.60	0.1131
y.l1	0.9469	0.0336	28.19	0.0000
trend	0.0003	0.0002	1.61	0.1105

Table 4.6. PP–test: Regression for consumption with constant and trend

The value of the $Z(t_{\tilde{\alpha}})$ statistic is -1.92, which is insignificant. The relevant Z statistics for the deterministic regressors are 0.72 and 2.57. Both of these are insignificant if compared with the critical values of Tables II and III in Dickey and Fuller [17] at the 5% level. These results are summarized in Table 4.7.

	test statistic	1%	5%	10%
$Z(t_{\hat{\alpha}})$	-1.92	-4.05	-3.46	-3.15
$Z(t_{\hat{\mu}})$	0.72	3.78	3.11	2.73
$Z(t_{\hat{\beta}})$	2.57	3.53	2.79	2.38

Table 4.7. PP–test: $Z(t_{\hat{\alpha}})$, $Z(t_{\hat{\mu}})$, and $Z(t_{\hat{\beta}})$ test

Next, step the trend is dropped from the test regression and the results are stored in the object lc.co (see command line 6). The regression results are summarized in Table 4.8. Again, the null hypothesis of a unit root cannot be rejected and the drift term is insignificant given the test statistics and critical values reported in Table 4.9. The critical values for the drift term now correspond to the ones provided for a sample size of 100 in Table I of Dickey and Fuller [17].

	Estimate	Std. Error	t value	Pr(>|t|)
(Intercept)	0.0109	0.0825	0.13	0.8954
y.l1	0.9996	0.0076	130.78	0.0000

Table 4.8. PP–test: Regression for consumption with constant only

So far, the conclusion about the integration order for the consumption series is the same as the one obtained by applying the sequential testing procedure of the ADF–test. Finally, it is checked whether differencing the series

	test statistic	1%	5%	10%
$Z(t_{\hat{\alpha}})$	−0.13	−3.50	−2.89	−2.58
$Z(t_{\hat{\mu}})$	0.20	3.22	2.54	2.17

Table 4.9. PP–test: $Z(t_{\hat{\alpha}})$ and $Z(t_{\hat{\mu}})$ test

once suffices to achieve stationarity (see command lines 7 and 8). The test regression is reported in Table 4.10. The values of the test statistic $Z(t_{\hat{\alpha}})$ is -10.96, which is highly significant, and therefore, it is concluded according to the results of the PP–test that the consumption series behaves like a pure random walk.

	Estimate	Std. Error	t value	Pr($>$|t|)
(Intercept)	0.0073	0.0016	4.73	0.0000
y.l1	−0.1253	0.1025	−1.22	0.2249
trend	0.0000	0.0000	0.36	0.7192

Table 4.10. PP–test: Regression for testing $I(2)$

4.3 ERS–Test

A shortcoming of the two previously introduced unit root tests is their low power if the true data-generating process is an autoregressive (AR)(1)–process with a coefficient close to one. To improve the power of the unit root test, Elliot *et al.* [24] proposed a local to unity detrending of the time series (henceforth: ERS–tests). The authors developed feasible point optimal tests, denoted as P_T^{μ} and P_T^{τ}, which take serial correlation of the error term into account. The second test type is denoted as the $DF - GLS$ test, which is a modified ADF–type test applied to the detrended data without intercept. The following model is entertained as the data-generating process for the series $y_1, \ldots, y_T$:

$$y_t = d_t + u_t , \tag{4.6a}$$

$$u_t = a u_{t-1} + v_t , \tag{4.6b}$$

where $d_t = \boldsymbol{\beta}' \boldsymbol{z}_t$ is a deterministic component; *i.e.*, a constant or a linear trend is included in the $(q \times 1)$ vector z_t, and v_t is a stationary zero-mean error process. In the case of $a = 1$, Equations 4.6a and 4.6b imply an integration order $I(1)$ for y_t, whereas $|a| < 1$ yields a stationary process for the series.

Let us first focus on the feasible point-optimal test statistic, which is defined as

$$P_T = \frac{S(a = \bar{a}) - \bar{a}S(a = 1)}{\hat{\omega}^2} , \qquad (4.7)$$

where $S(a = \bar{a})$ and $S(a = 1)$ are the sum of squared residuals from a least-squares regression of y_a on Z_a with

$$y_a = (y_1, y_2 - ay_1, \ldots, y_T - ay_{T-1}) , \qquad (4.8a)$$
$$Z_a = (z_1, z_2 - az_1, \ldots, z_T - az_{T-1}) ; \qquad (4.8b)$$

hence, y_a is a T-dimensional column vector and Z_a defines a $(T \times q)$ matrix. The estimator for the variance of the error process v_t can be estimated with

$$\hat{\omega} = \frac{\hat{\sigma}_\nu^2}{(1 - \sum_{i=1}^{p} \hat{\alpha}_i)^2} , \qquad (4.9)$$

where $\hat{\sigma}_\nu^2$ and $\hat{\alpha}_i$ for $i = 1, \ldots, p$ are taken from the auxilliary ordinary least-squares (OLS) regression:

$$\Delta y_t = \alpha_0 + \alpha_1 \Delta y_{t-1} + \ldots + \Delta y_{t-p} + \alpha_{p+1} + \nu_t . \qquad (4.10)$$

Finally, the scalar $\bar{a}$ is set to $\bar{a} = 1 + \bar{c}/T$, where $\bar{c}$ denotes a constant. Depending on the deterministic components in z_t, $\bar{c}$ is set either to -7 in the case of a constant or to -13.5 in the case of a linear trend. These values have been derived from the asymptotic power functions and its enevlope. Critical values of the P_T^μ and P_T^τ tests are provided in Table I of Elliott *et al.* [24].

Next, the authors propose a modified ADF–type test, which is the t statistic for testing $\alpha_0 = 0$ in the homogenous regression:

$$\Delta y_t^d = \alpha_0 y_{t-1}^d + \alpha_1 \Delta y_{t-1}^d + \ldots + \alpha_p \Delta y_{t-p}^d + \varepsilon_t , \qquad (4.11)$$

where y_t^d are the residuals in the auxilliary regression $y_t^d \equiv y_t - \hat{\beta}' z_t$. When there is no intercept one can apply the critical values of the typical DF–type t tests; in the other instances critical values are provided in Table I of Elliott *et al.* [24].

Both test types have been implemented as function `ur.ers()` in the contributed package `urca`. The function allows the provision of the number of lags to be included in the test regression for the $DF - GLS$ test via its argument `lag.max`. The optimal number of lags for estimating $\hat{\omega}$ is determined by the Bayesian information criterion (BIC). A summary method for objects of class `ur.ers` exists that return the test regression in the case of the $DF - GLS$ test and the value of the test statistic with the relevant critical values for the 1%, 5%, and 10% significance level for both tests. In the Rcode example 4.3, both test types are applied to the logarithm of real Gross National Product (GNP) used in the seminal paper of Nelson and Plosser [69].

Rcode 4.3 ERS–tests: Integration order for real GNP in the United States

```
library(urca)                                                      1
data(nporg)                                                        2
gnp <- log(na.omit(nporg[, "gnp.r"]))                              3
gnp.d <- diff(gnp)                                                 4
gnp.ct.df <- ur.ers(gnp, type = "DF-GLS", model = "trend",        5
    lag.max = 4)
gnp.ct.pt <- ur.ers(gnp, type = "P-test", model = "trend")        6
gnp.d.ct.df <- ur.ers(gnp.d, type = "DF-GLS", model = "trend      7
    ", lag.max = 4)
gnp.d.ct.pt <- ur.ers(gnp.d, type = "P-test", model = "trend      8
    ")
```

First, the P_T^τ and $DF - GLS^\tau$ are applied to the series, where in the case of the $DF - GLS^\tau$, four lags have been added (see command lines 5 and 6). The test results are displayed in Table 4.11. Both tests imply a unit root for the data-generating process.

	test statistic	1%	5%	10%
P_T^τ	6.65	4.26	5.64	6.79
$DF - GLS^\tau$	-2.08	-3.58	-3.03	-2.74

Table 4.11. ERS–tests: P_T^τ and $DF - GLS^\tau$ for real GNP of the United States

In a second step, both tests are applied to the differenced series. As the results summarized in Table 4.12 imply, differencing the series once suffices to achieve stationarity.

	test statistic	1%	5%	10%
P_T^τ	2.68	4.26	5.64	6.79
$DF - GLS^\tau$	-4.13	-3.58	-3.03	-2.74

Table 4.12. ERS–tests: P_T^τ and $DF - GLS^\tau$ for testing $I(2)$

4.4 Schmidt–Phillips (SP)–Test

In Section 4.3, unit root tests have been described that are more powerful than the usual DF–type tests. Another drawback of the DF–type tests is that the nuisance parameters, *i.e.*, the coefficients of the deterministic regressors, are either not defined or have a different interpretation under the alternative hypothesis of stationarity. To elucidate this point, consider the test regressions as in Equations 4.1a-4.1c again. Equation 4.1c does neither allow a nonzero level nor a trend under both the null and the alternative hypothesis. Whereas in Equations 4.1a and 4.1b, these regressors are taken into account, but now these coefficients have a different interpretation under the null and the alternative. The constant term β_1 in Equation 4.1b has the interpretation of a deterministic trend under the null hypothesis; $i.e$, $\pi = 1$, but it has to be considered as a level regressor under the alternative. Likewise, in Equation 4.1a, β_1 represents a linear trend and β_2 represents a quadratic trend under the null hypothesis of integratedness, but these coefficients have the interpretation of a level and linear trend regressor under the alternative hypothesis of stationarity. Schmidt and Phillips [88] proposed a Lagrange multiplier (LM)-type test statistic that defines the same set of nuisance parameters under both the null and the alternative hypothesis. Furthermore, they consider higher polynomials than a linear trend. The authors consider the following model:

$$y_t = \alpha + \mathbf{Z}_t\boldsymbol{\delta} + x_t , \tag{4.12a}$$

$$x_t = \pi x_{t-1} + \varepsilon_t , \tag{4.12b}$$

where ε_t are i.i.d $\mathcal{N}(0, \sigma^2)$ and $\mathbf{Z}_t = (t, t^2, \ldots, t^p)$. The test statistic $\tilde{\rho}$ is then constructed by running the regression:

$$\Delta y_t = \Delta \mathbf{Z}_t\boldsymbol{\delta} + u_t , \tag{4.13}$$

first and calculating $\tilde{\psi}_x = y_1 - \mathbf{Z}_1\tilde{\boldsymbol{\delta}}$, where $\tilde{\boldsymbol{\delta}}$ is the OLS estimate of δ in Equation 4.13. Next a series $\tilde{S}_t$ is defined as $\tilde{S}_t = y_t - \tilde{\psi}_x - \mathbf{Z}_t\tilde{\boldsymbol{\delta}}$. Finally, the test regression is then given by

$$\Delta y_t = \Delta \mathbf{Z}_t\boldsymbol{\gamma} + \phi \tilde{S}_{t-1} + v_t , \tag{4.14}$$

where v_t assigns an error term. The test statistic is then defined as $Z(\rho) = \frac{\tilde{\rho}}{\hat{\omega}^2} = \frac{T\tilde{\phi}}{\hat{\omega}^2}$ with $\tilde{\phi}$ as the OLS estimate of ϕ in Equation 4.14, and an estimator for ω^2 is given by

$$\hat{\omega}^2 = \frac{T^{-1}\sum_{i=1}^{T}\hat{\varepsilon}_t^2}{T^{-1}\sum_{i=1}^{T}\hat{\varepsilon}_t^2 + 2T^{-1}\sum_{s=1}^{l}\sum_{t=s+1}^{T}\hat{\varepsilon}_t\hat{\varepsilon}_{t-s}} , \tag{4.15}$$

where $\hat{\varepsilon}_t$ are the residuals from Equation 4.12. Depending on the sample size and the order of the polynomial $\mathbf{Z}$, critical values are provided in Schmidt and Phillips [88]. Aside from this test statistic, one can also apply the t ratio statistic $Z(\tau) = \frac{\tilde{\tau}}{\hat{\omega}^2}$ for testing $\phi = 0$. As shown by a Monte Carlo simulation, these tests fair better in terms of power compared with the corresponding DF–type test statistic.

These two tests have been implemented as function **ur.sp()** in the contributed package **urca**. As arguments the series name, the test type, either **tau** for $\tilde{\tau}$ or **rho** for $\tilde{\rho}$, the polynomial degree and the signficance level have to be entered in **ur.sp()**. In Rcode example 4.4, these tests have been applied to the nominal GNP series of the United States expressed in millions of current U.S. dollars as used by Nelson and Plosser [69]. By eyeball inspection of the series as displayed in Figure 4.2, a quadratic trend is assumed.

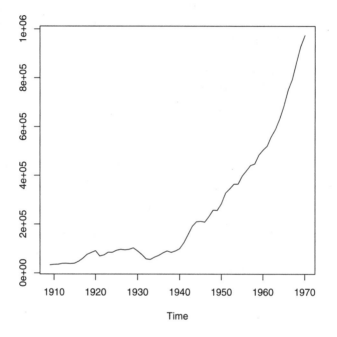

Fig. 4.2. Nominal GNP of the United States

Rcode 4.4 SP–test: Integration order for nominal GNP in the United States

```
library(urca)                                                    1
data(nporg)                                                      2
gnp <- na.omit(nporg[, "gnp.n"])                                 3
gnp.tau.sp <- ur.sp(gnp, type = "tau", pol.deg=2, signif         4
    =0.05)
gnp.rho.sp <- ur.sp(gnp, type = "rho", pol.deg=2, signif         5
    =0.05)
```

	Estimate	Std. Error	t value	Pr(>|t|)
(Intercept)	9355.4016	7395.3026	1.27	0.2110
y.lagged	0.9881	0.0411	24.03	0.0000
trend.exp1	−982.3230	610.1348	−1.61	0.1129
trend.exp2	30.2762	16.1573	1.87	0.0661

Table 4.13. SP–test: Result of level regression with polynomial of order two

This setting is evidenced by a significant coefficient of the quadratic trend regressor in the unconstrained model 4.12 as displayed in Table 4.13. The results of the two tests are displayed in Table 4.14. Both tests indicate an integration order of $I(1)$ at least for a significance level of 5%.

	test statistic	1%	5%	10%
$\tilde{\tau}$	−2.03	−4.16	−3.65	−3.34
$\tilde{\rho}$	−8.51	−30.40	−23.70	−20.40

Table 4.14. SP–tests: $\tilde{\tau}$ and $\tilde{\rho}$ for nominal GNP of the United States

4.5 KPSS–Test

Kwiatkowski *et al.* [56] proposed a LM-test for testing trend and/or level stationarity (henceforth: KPSS–test). That is, now the null hypothesis is a stationary process, whereas in the former tests, it was a unit root process. By taking the null hypothesis as a stationary process and the unit root as an alternative is in accordance with a conservative testing strategy. One should always seek for tests that place the hypothesis we are interested in as the alternative one. Hence, if we then reject the null hypothesis, we can be pretty confident that the series has indeed a unit root. Therefore, if the results of the above tests do indicate a unit root, but the result of the KPSS–test does indicate a stationary process, one should be cautious and opt for the latter result. They consider the following model:

$$y_t = \xi t + r_t + \varepsilon_t , \qquad (4.16a)$$

$$r_t = r_{t-1} + u_t , \qquad (4.16b)$$

where r_t is a random walk and the error process is assumed to be i.i.d $(0, \sigma_u^2)$. The initial value r_0 is fixed and corresponds to the level. If $\xi = 0$, then this model is in terms of a constant only as deterministic regressors. Under the null hypothesis, ε_t is stationary and therefore y_t is either trend stationary or in the case of $\xi = 0$, level stationary. The test statistic is constructed as follows:

First, regress y_t on a constant or on a constant and a trend depending on whether one wants to test level- or trend-stationarity; second, calculate the partial sums of the residuals $\hat{\varepsilon}_t$ from this regression as

$$S_t = \sum_{i=1}^{t} \hat{\varepsilon}_i \, , t = 1, 2, \ldots, T \, . \tag{4.17}$$

The test statistic is then defined as

$$LM = \frac{\sum_{t=1}^{T} S_t^2}{\hat{\sigma}_\varepsilon^2} \, , \tag{4.18}$$

with $\hat{\sigma}_\varepsilon^2$ being an estimate of the error variance from step one. The authors suggest the utilization of a Bartlett window $w(s,l) = 1 - s/(l+1)$ as an optimal weighting function to estimate the long-run variance $\hat{\sigma}_\varepsilon^2$; that is

$$\hat{\sigma}_\varepsilon^2 = s^2(l) = T^{-1} \sum_{t=1}^{T} \hat{\varepsilon}_t^2 + 2T^{-1} \sum_{s=1}^{l} 1 - \frac{s}{l+1} \sum_{t=s+1}^{T} \hat{\varepsilon}_t \hat{\varepsilon}_{t-1} \, . \tag{4.19}$$

The upper tail critical values of the level- and trend-stationarity version are given in Kwiatkowski *et al.* [56].

The two test types are implemented as function ur.kpss() in the contributed package urca and as function kpss.test() in the contributed package tseries.[3] The implementation as in the package urca will be applied to the Nelson and Plosser [69] data set and thereby replicate some results in Kwiatkowski *et al.* [56]. Aside from the test type mu or tau for level-stationarity or trend-stationarity, the user has the option to set the Bartlett window parameter l *via* the argument lags either to short, which corresponds to the integer value of $4 \times (T/100)^{1/4}$, or to long, which is equivalent to the integer value of $12 \times (T/100)^{1/4}$ or to nil. Alternatively, one can specify an integer value by providing the desired lag length *via* the argument use.lag. In Rcode example 4.5, the level-stationary version is applied to the interest rate data and the trend-stationary version of the test is applied to the logarithm of nominal wages. A lag length l of eight has been used by setting the functional argument use.lag accordingly.

---

[3] Both functions have been ported into the contributed package fSeries as urkpssTest() and tskpssTest(), respectively.

Rcode 4.5 KPSS–test: Integration order for interest rate and nominal wages in the United States

```
library(urca)                                              1
data(nporg)                                                2
ir <- na.omit(nporg[, "bnd"])                              3
wg <- log(na.omit(nporg[, "wg.n"]))                        4
ir.kpss <- ur.kpss(ir, type = "mu", use.lag=8)             5
wg.kpss <- ur.kpss(wg, type = "tau", use.lag=8)            6
```

The null hypothesis of level- and trend-stationarity, respectively, cannot be rejected for both series as shown in Table 4.15.

	test statistic	1%	5%	10%
$\hat{\eta}_\mu$	0.13	0.35	0.46	0.74
$\hat{\eta}_\tau$	0.10	0.12	0.15	0.22

Table 4.15. KPSS–tests: $\hat{\eta}_\mu$ and $\hat{\eta}_\tau$ for interest rates and nominal wages of the United States

Summary

In this first chapter of Part II, various unit root tests have been applied to real data sets. The sequential testing strategy of the ADF–test outlined in Section 2.2 has been applied to U.K. consumption. Because the data-generating process is unknown, it is recommended to go through these steps rather than merely apply one test regressions as in Equations 4.1a-4.1c. Furthermore, a spherical error term should always be ensured by supplying sufficient lagged endogenous variables. Next, the Phillips and Perron–test has been applied to the same data set. In principle, the difference between the two tests is that the latter uses a nonparametric correction that captures weak dependence and heterogenity of the error process. As has been pointed out in Section 4.3, the relatively low power of both tests due to the fact that a unit root process is specified as the null hypothesis must be considered as a shortcoming. The ERS–tests ameliorate this problem and should therefore be preferred. Furthermore, the nuisance parameters have different interpretation if either the null or the alternative hypothesis is true. The SP–test adresses this problem explicitly and allows for inclusion of higher polynomials in the deterministic part. However, all tests suffer from an ill-specified null hypothesis. The KPSS–test, as a test for stationarity, adresses the hypothesis specification from the viewpoint of a conservative testing correctly. Anyway, unfortunately there is no clear-cut answer to the question of which test should be applied to a data set. A combination of some of the above-mentioned tests with the inclusion of opposing null hypothesis seems, therefore, to be a pragmatic approach in practice.

Testing for the presence of seasonal unit roots as it was touched on in Section 2.2 will be a topic in the next chapter as well as how the occurence of structural breaks will affect unit root inference.

Exercises

1. Determine the order of integration for the income and wealth series contained in the data set `Raotbl3` with the ADF– and the PP–test.

2. Apply the ERS–tests to the Nelson and Plosser data set contained in `nporg`, and compare your findings with the ones in Nelson and Plosser [69].

3. Replicate the results in Kwiatkowski *et al.* [56], and compare again with the results in Nelson and Plosser [69].

4. Response surface regression for the ERS–tests P_T^μ and P_T^τ:

   a) First, write a function that displays the critical values of the P_T^μ and P_T^τ statistic as provided in Table I of Elliott *et al.* [24].

   b) Next, write a function for conducting a Monte Carlo simulation of the P_T^μ P_T^τ statistic for finite samples.

   c) Fit a response surface regression to your results from exercise 4b.

   d) Finally, compare the critical values implied from the response surface regression with the ones provided in Table I of Elliott *et al.* [24] for selected significance levels and sample sizes.

5. Complete the following table:

data set	ADF	PP	ERS	SP	KPSS
(Raotbl3) lc	$I(1)$	$I(1)$			
(nporg) gnp.r			$I(1)$		
(nporg) gnp.n				$I(1)$	
(nporg) bnd					$I(0)$
(nporg) wg.n					$I(0)$

5

Further Considerations

In Chapter 4, various unit root tests were introduced and compared among each other. This chapter deals with two further topics. First, the case of structural breaks in a time series is considered and how this affects the inference about the degree of integratedness. Second, the issue of seasonal unit roots is discussed as it was only briefly touched in Section 2.2.

5.1 Stable Autoregressive (AR)(1)–Processes with Structural Breaks

Recall from Section 2.2 the random walk with drift model as in Equation 2.7. It has been argued that the drift parameter μ can be viewed as a deterministic trend, given the final form as

$$y_t = \mu t + y_0 + \sum_{i=1}^{t} \varepsilon_t \; . \tag{5.1}$$

Now suppose that the series is contaminated by a structural break. Such an occurence can be caused by a new legislation that affects the economy or can be caused by a redefintion of the data series; *e.g.*, a new definition for counting the unemployed has been decreed. One can distuinguish two different ways for how such a structural shift impacts a series. Either the break occurs at only one point in time and then lasts for the remaing periods of the sample or it influences the series only in one particular period. In practice, such structural shifts are modeled by introducing dummy variables. In the former case, the structural shift is modeled as a step dummy variable that is zero before the break date and unity afterward. The latter is referred to as a pulse intervention, and the dummy variable is only unity at the break date and zero otherwise. Either way, if the series is $I(1)$, such a structural shift will have a lasting effect on the series. Consider the following data-generated process:

$$y_t = \mu + \delta D_t + y_{t-1} + \varepsilon_t \; , \tag{5.2}$$

where D_t assigns a pulse dummy variable that is defined as

$$D_t = \begin{cases} 1 & t = \tau \; , \\ 0 & \text{otherwise} \; , \end{cases}$$

where τ assigns the break date. Even though the break occurs only in one period, it will have a lasting effect on the series as can be seen by calculating the final form of Equation 5.2, which is given by

$$y_t = \mu + \delta S_t + y_0 + \sum_{i=1}^{t} \varepsilon_t \ , \tag{5.3}$$

and S_t is

$$S_t = \begin{cases} 1 & t \geq \tau \ , \\ 0 & \text{otherwise} \ . \end{cases}$$

In Rcode example 5.1, two random walks with drift have been generated from the same sequence of disturbances with size 500. The second process has been affected at observation 250 with a pulse dummy, defined as object S (see command lines 5 and 9). The two series are plotted in Figure 5.1.

Rcode 5.1 Random walk with drift and structural break

```
set.seed(123456)                                              1
e <- rnorm(500)                                               2
# trend                                                       3
trd <- 1:500                                                  4
S <- c(rep(0, 249), rep(1, 251))                              5
# random walk with drift                                      6
y1 <- 0.1*trd + cumsum(e)                                     7
# random walk with drift and shift                            8
y2 <- 0.1*trd + 10*S + cumsum(e)                              9
# plotting                                                    10
par(mar=rep(5,4))                                             11
plot.ts(y1, lty=1, ylab='', xlab='')                         12
lines(y2, lty=2)                                              13
legend(10, 50, legend=c('rw with drift', 'rw with drift and  14
    pulse'), lty=c(1, 2))
```

The difficulty in statistically distuinguishing an $I(1)$–series from a stable $I(0)$ one develops if the latter is contaminated by a structrual shift. Hence, the inference drawn from a Dickey–Fuller (DF)–type test becomes unreliable in the case of a potential structural break. It has been shown by Perron [75][76][77] and Perron and Vogelsang [78]. In Perron [75], the author considers three different kinds of models, where the structural break point is assumed to be known:

Model (A): $y_t = \mu + dD(T_\tau) + y_{t-1} + \varepsilon_t \ , \tag{5.4a}$

Model (B): $y_t = \mu_1 + (\mu_2 - \mu_1)DU_t + y_{t-1} + \varepsilon_t \ , \tag{5.4b}$

Model (C): $y_t = \mu_1 + dD(T_\tau) + (\mu_2 - \mu_1)DU_t + y_{t-1} + \varepsilon_t \ , \tag{5.4c}$

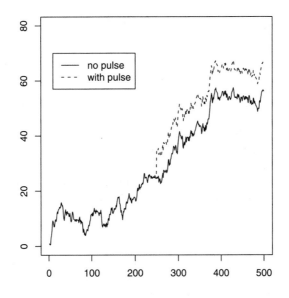

Fig. 5.1. Time series plot of random walk with drift and structural break

where $1 < T_\tau < T$ assigns the *a priori* known break point; $D(T_\tau) = 1$ if $t = T_\tau + 1$ and 0 otherwise; $DU_t = 1$ for $t > T_\tau$ and 0 otherwise. It is further assumed that the error process can be represented as $\phi(L)\varepsilon_t = \theta(L)\xi_t$ with ξ i.i.d., where $\phi(L)$ and $\theta(L)$ assign lag polynomials. In model (A), a one-time shift in the levels of the series is taken into account, whereas in model (B), a change in the rate of growth is allowed and model (C) is a combination of both. The model specifications for the trend-stationary alternative are

Model (A): $y_t = \mu_1 + \beta t + (\mu_2 - \mu_1)DU_t + \varepsilon_t$,   (5.5a)

Model (B): $y_t = \mu + \beta_1 t + (\beta_2 - \beta_1)DT_t^* + \varepsilon_t$,   (5.5b)

Model (C): $y_t = \mu + \beta_1 t + (\mu_2 - \mu_1)DU_t + (\beta_2 - \beta_1)DT_t^* + \varepsilon_t$,   (5.5c)

where $DT_t^* = t - T_\tau$ for $t > T_\tau$ and 0 otherwise.

The author proposed an adjusted, augmented Dickey–Fuller (ADF)–type test for the three models, which is based on the following test regressions:

$$y_t = \hat{\mu}^A + \hat{\theta}^A DU_t + \hat{\beta}^A t + \hat{d}^A D(T_\tau)_t + \hat{\alpha}^A y_{t-1} + \sum_{i=1}^{k} \hat{c}_i^A \Delta y_{t-i} + \hat{\varepsilon}_t , \quad (5.6a)$$

$$y_t = \hat{\mu}^B + \hat{\beta}^B t + \hat{\gamma}^B DT_t^* + \hat{\alpha}^B y_{t-1} + \sum_{i=1}^{k} \hat{c}_i^B \Delta y_{t-i} + \hat{\varepsilon}_t , \quad (5.6b)$$

$$y_t = \hat{\mu}^C + \hat{\theta}^C DU_t + \hat{\beta}^C t + \hat{\gamma}^C DT_t^* + \hat{d}^C D(T_\tau)_t + \hat{\alpha}^C y_{t-1}$$
$$+ \sum_{i=1}^{k} \hat{c}_i^C \Delta y_{t-i} + \hat{\varepsilon}_t . \quad (5.6c)$$

The test statistic is the Student t ratio $t_{\hat{\alpha}^i}(\lambda)$ for $i = A, B, C$. Please note that this test statistic is now dependent on the fraction of the structural break point with respect to the total sample, i.e, $\lambda = \frac{T_\tau}{T}$. The critical values of this test statistic are provided in Perron [75][77]. The author applied these models to the data series used in Nelson and Plosser [69] and concluded that most of the series does not contain any longer a unit root if the presence of a structural break is taken into account.

Zivot and Andrews [107] pointed out that the risk of data mining exists if the break point is set exogenously by the researcher. They propose a test that circumvents this possibility by endogenously determining the most likely occurence of a structural shift. By reanalyzing the data set used in Perron [75], they found less evidence for rejecting the assumption of a unit root process. The estimation procedure proposed by them is to choose the date of the structural shift for that point in time that gives the least favorable result for the null hypothesis of a random walk with drift. The test statistic is likewise to Perron [75] the Student t ratio:

$$t_{\hat{\alpha}^i}[\hat{\lambda}_{\text{inf}}^i] = \inf_{\lambda \in \Delta} t_{\hat{\alpha}^i}(\lambda) \quad \text{for} \quad i = A, B, C , \quad (5.7)$$

where Δ is a closed subset of $(0, 1)$. Depending on which model is selected, the test statistic is inferred from one of the following test regressions:

$$y_t = \hat{\mu}^A + \hat{\theta}^A DU_t(\hat{\lambda}) + \hat{\beta}^A t + \hat{\alpha}^A y_{t-1} + \sum i = 1^k \hat{c}_i^A \Delta y_{t-i} + \hat{\varepsilon}_t , \quad (5.8a)$$

$$y_t = \hat{\mu}^B + \hat{\beta}^B t + \hat{\gamma}^B DT_t^*(\hat{\lambda}) + \hat{\alpha}^B y_{t-1} + \sum i = 1^k \hat{c}_i^B \Delta y_{t-i} + \hat{\varepsilon}_t , \quad (5.8b)$$

$$y_t = \hat{\mu}^C + \hat{\theta}^C DU_t(\hat{\lambda}) + \hat{\beta}^C t + \hat{\gamma}^C DT_t^*(\hat{\lambda}) + \hat{\alpha}^C y_{t-1}$$
$$+ \sum i = 1^k \hat{c}_i^C \Delta y_{t-i} + \hat{\varepsilon}_t , \quad (5.8c)$$

where $DU_t(\lambda) = 1$ if $t > T\lambda$ and 0 otherwise, and $DT_t^*(\lambda) = t - T\lambda$ for $t > T\lambda$ and 0 otherwise. Because now λ is estimated, one cannot any longer use the critical values as in Perron [75][77], but the values published in Zivot and Andrews [107] have to be used instead.

The Zivot and Andrews test is implemented as function ur.za() in the contributed package urca. In Rcode example 5.2, this test is applied to the nominal and real wage series of the Nelson and Plosser data set. The test is applied to the natural logarithm of the two series. With the functional argument model, the type can be specified in which intercept stands for model specification (A), trend corresponds to model type (B) and both is model type (C). The integer value of lag determines the number of lagged endogenous variables to be included in the test regression.

Rcode 5.2 Unit roots and structural break: Zivot–Andrews test

```
library(urca)                                              1
data(nporg)                                                2
wg.n <- log(na.omit(nporg[, "wg.n"]))                      3
za.wg.n <- ur.za(wg.n, model = "intercept", lag = 7)       4
plot(za.wg.n)                                              5
wg.r <- log(na.omit(nporg[, "wg.r"]))                      6
za.wg.r <- ur.za(wg.r, model = "both", lag = 8)            7
plot(za.wg.r)                                              8
```

The regression output, *i.e.*, the contents of slot testreg, is displayed in Tables 5.1 and 5.2, respectively.

	Estimate	Std. Error	t value	Pr(>|t|)
(Intercept)	1.9878	0.3724	5.34	0.0000
y.l1	0.6600	0.0641	10.30	0.0000
trend	0.0173	0.0033	5.32	0.0000
y.dl1	0.4979	0.1121	4.44	0.0000
y.dl2	0.0557	0.1308	0.43	0.6717
y.dl3	0.1494	0.1278	1.17	0.2477
y.dl4	0.0611	0.1266	0.48	0.6314
y.dl5	0.0061	0.1264	0.05	0.9616
y.dl6	0.1419	0.1249	1.14	0.2610
y.dl7	0.2671	0.1195	2.24	0.0297
du	−0.1608	0.0387	−4.16	0.0001

Table 5.1. Zivot–Andrews: Test regression for nominal wages

The results of the test statistic are provided in Table 5.3. The unit root hypothesis must be rejected for the nominal wage series, given a significance level of 5%, whereas the unit root hypothesis cannot be rejected for the real wage series. The structural shift for the nominal wage series occurred most likely in period 30, which corresponds to the year 1929. The estimated break

	Estimate	Std. Error	t value	Pr(>|t|)
(Intercept)	2.5671	0.5327	4.82	0.0000
y.l1	0.1146	0.1866	0.61	0.5420
trend	0.0124	0.0028	4.49	0.0000
y.dl1	0.6111	0.1662	3.68	0.0006
y.dl2	0.3516	0.1686	2.09	0.0423
y.dl3	0.4413	0.1568	2.82	0.0070
y.dl4	0.2564	0.1453	1.76	0.0838
y.dl5	0.1381	0.1346	1.03	0.3100
y.dl6	0.0591	0.1262	0.47	0.6416
y.dl7	0.1673	0.1201	1.39	0.1697
y.dl8	0.1486	0.1210	1.23	0.2254
du	0.0849	0.0196	4.33	0.0001
dt	0.0081	0.0022	3.68	0.0006

Table 5.2. Zivot–Andrews: Test regression for real wages

point is stored in the slot `bpoint`.

	test statistic	1%	5%	10%
wages, nominal	−5.30	−5.34	−4.80	−4.58
wages, real	−4.74	−5.57	−5.08	−4.82

Table 5.3. Zivot–Andrews: Test statistics for real and nominal wages

Finally, beside a summary method for objects of class `ur.za`, a plot method exists that depicts the path of the test statistic. The significance levels are drawn as separate lines, and in the case of a structural break, the break point is highlighted by a dashed vertical line. The graphs are displayed in Figures 5.2 and 5.3, respectively.

5.2 Seasonal Unit Roots

In Section 2.2, the topic of seasonal unit roots has been briefly discussed. We will now investigate the issue of seasonal integration more thoroughly. This need originates because often applied economists need to construct models for seasonally unadjusted data. The reason for this is twofold. First, some data might be obviously seasonally in nature, and second, sometimes the utilization of seasonally adjusted data might distort the dynamics of an estimated model, as has been pointed out by Wallis [99].

Zivot and Andrews Unit Root Test

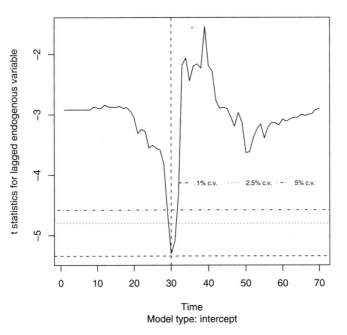

Fig. 5.2. Zivot–Andrews test statistic for nominal wages

Recall the seasonal difference operator and its factorization as shown in Equations 2.10a and 2.10b. For quarterly data, this factorization would yield

$$(1 - L^4) = (1 - L)(1 + L)(1 - iL)(1 + iL), \qquad (5.9)$$

where $\pm i$ are complex roots.[1] A seasonal quarterly process has therefore four possible roots, namely 1, −1, and $\pm i$. These roots correspond to different cycles in the time domain. The root 1 has a single period cycle and is the zero frequency root. The root −1 has a two period cycle that implies for quarterly data a biannual cycle. Finally, the complex roots have a cycle of four periods that is equivalent to one cycle per year in quarterly data. The problem caused by the complex roots for quarterly data is that the effects of these are indistinguishable from each other. In Table 5.2, the cycles by the roots of the seasonal difference operator are summarized.

As mentioned in Section 2.2, the first attempts to test for seasonal unit roots and probably the most simplest have been suggested by Hasza and

---

[1] For brevity, we consider quarterly data only. The factorizations for the other seasonal frequencies are provided in Franses and Hobijn [30], for example.

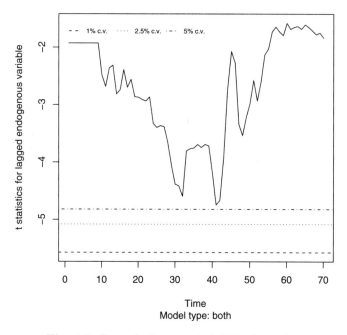

Fig. 5.3. Zivot–Andrews test statistic for real wages

Root $+1$	Root -1	Root $+i$	Root $-i$
Factor $(1-L)$	Factor $(1+L)$	Factor $(1-iL)$	Factor $(1+iL)$
$y_{t+1} = y_t$	$y_{t+1} = -y_t$	$y_{t+1} = iy_t$	$y_{t+1} = -iy_t$
	$y_{t+2} = -(y_{t+1}) = y_t$	$y_{t+2} = i(y_{t+1}) = -y_t$	$y_{t+2} = -i(y_{t+1}) = -y_t$
		$y_{t+3} = i(y_{t+2}) = -iy_t$	$y_{t+3} = -i(y_{t+2}) = iy_t$
		$y_{t+4} = i(y_{t+3}) = y_t$	$y_{t+4} = -i(y_{t+3}) = y_t$

Table 5.4. Cycles implied by the roots of the seasonal difference operator

Fuller [40] and Dickey *et al.* [18]. The latter authors suggested the following test regression:

$$\Delta_s z_t = \delta_0 z_{t-1} + \sum_{i=1}^{k} \delta_i \Delta_s y_{t-i} + \varepsilon_t . \tag{5.10}$$

The variable z_t is constructed by estimating the auxilliary regression:

$$\Delta_s y_t = \sum_{i=1}^{h} \lambda_i \Delta_s y_{t-i} + \varepsilon_t , \tag{5.11}$$

which yields the coefficient estimates $\hat{\lambda}_1, \dots, \hat{\lambda}_h$. The variable z_t is then constructed as

$$z_t = y_t - \sum_{i=1}^{h} \hat{\lambda}_i y_{t-i} \, . \tag{5.12}$$

The test for seasonal integration is then based on the Student t ratio for the ordinary least-squares estimate of the coefficient δ_0 in Equation 5.10. Osborn et al. [71] suggested to replace $\Delta_s z_t$ with $\Delta_s y_t$ as the dependent variable in Equation 5.10. Incidentally, if $h = 0$, this is equivalent with an ADF–regression for the seasonal differences; i.e.,

$$\Delta_s y_t = \delta_0 y_{t-s} + \sum_{i=1}^{k} \delta_i \Delta_s y_{t-i} + \varepsilon_t \, . \tag{5.13}$$

The lag order k and h should be determined similar to the procedures proposed for the ADF–test in Section 2.2. Furthermore, it should be noted that deterministic seasonal dummy variables can also be included in the test regression. The relevant critical values are provided in Osborn et al. [71] and are dependent on the inclusion of such deterministic dummy variables and whether the data have been demeaned at the seasonal frequency. If the null hypothesis of the existence of a seasonal unit root is rejected for an absolute large enough t ratio, then one might conclude that stochastic seasonality is not present or that stochastic seasonality, which can be removed by using s-differences, does not exist. On the other side, if the null hypothesis cannot be rejected it is common practice to consider the order of nonseasonal differencing required to achieve stationarity instead of considering higher orders of seasonal differencing. Hence, one might consider as a data-generating process $SI(0,0)$ or $SI(0,1)$ or $SI(d,1)$ at most. To discriminate between $SI(0,1)$ and $SI(1,1)$ where the former is the new null hypothesis, one can do so by estimating the following ADF–type test regression:

$$\Delta\Delta_s y_t = \delta_0 \Delta_s y_{t-1} + \sum_{i=1}^{k} \delta_i \Delta\Delta_s y_{t-i} + \varepsilon_t \tag{5.14}$$

and consider the t ratio for the hypothesis $\delta_0 = 0$. If this test statistic is insignificant, one takes $SI(2,1)$ as the new null hypothesis into account by estimating

$$\Delta\Delta\Delta_s y_t = \delta_0 \Delta\Delta_s y_{t-1} + \sum_{i=1}^{k} \delta_i \Delta\Delta\Delta_s y_{t-i} + \varepsilon_t \, , \tag{5.15}$$

and so forth.

The deficiency of the test regression proposed by Osborn et al. [71] is that it does not test all possible unit roots in a seasonal process (see Table 5.2). Hylleberg et al. [49] suggested a test that allows for cyclical movements at different frequencies and takes the factorization of the seasonal lag polynomial as in Table 5.2 explicitly into account (henceforth: HEGY–test). For quarterly data, they propose the test regression:

$$\Delta_4 y_t = \sum_{i=1}^{4} \pi_i Y_{i,t-i} + \varepsilon_t , \qquad (5.16)$$

where the regressors $Y_{i,t}$ for $i = 1, \ldots, 4$ are constructed as

$$Y_{1,t} = (1 + L)(1 + L^2)y_t = y_t + y_{t-1} + y_{t-2} + y_{t-3} , \qquad (5.17a)$$

$$Y_{2,t} = -(1 - L)(1 + L^2)y_t = -y_t + y_{t-1} - y_{t-2} + y_{t+3} , \qquad (5.17b)$$

$$Y_{3,t} = -(1 - L)(1 + L)y_t = -y_t + y_{t+2} , \qquad (5.17c)$$

$$Y_{4,t} = -(L)(1 - L)(1 + L)y_t = Y_{3,t-1} = -y_{t-1} + y_{t-3} . \qquad (5.17d)$$

The null hypothesis of seasonal integration implies that the coefficients π_i for $i = 1, \ldots, 4$ are equal to zero. As outlined, each π_i has a different interpretation. If only π_1 is significantly negative, then there is no nonseasonal stochastic stationary component in the data-generating process. Likewise, if only π_2 is significant, then there is no evidence of a biannual cycle in the data. Finally, the significance of π_3 and π_4 can be tested jointly with a Lagrange-multiplier F test. To put it differently, the existence of unit roots at the zero, biannual and annual frequency correspond to $\pi_1 = 0$, $\pi_2 = 0$, and $\pi_3 = \pi_4 = 0$, respectively. It should be noted that deterministic terms, like an intercept, a trend, seasonal dummy variables, or a combination of them, as well as lagged seasonal differences can be added to the test regression 5.16. That is, the general test regression is given by

$$\Delta_4 y_t = \pi_0 + \sum_{i=1}^{4} \pi_i Y_{i,t-1} + \sum_{i=1}^{3} \beta_i DS_{i,t} + \sum_{i=1}^{k} \delta_i \Delta_4 y_{t-1} + \gamma t + \varepsilon_t , \qquad (5.18)$$

where $DS_{i,t}$ assign the seasonal dummy variables, π_0 the constant term, and t a time trend. The critical values are provided in Hylleberg et al. [49] (HEGY) and are dependent on the specification choosen and the sample size.

The HEGY–test is implemented as function `HEGY.test` in the contributed package `uroot`.[2]

---

[2] Incidentally, the package is shipped with a graphical user interface that is launched by executing `urootgui()` from the console. Beside the HEGY–function, it should be noted at this point that the ADF–test with the option to include deterministic seasonal dummy variables is available as function `ADF.test()` as well as the tests proposed by Canova and Hansen [10] as functions `CH.test()` and textttCH-seas.test(), respectively, which is a generalization of the KPSS–test from the zero frequency to the seasonal frequencies. Other features of the package are the generation of a LaTeX table containing the test results, a panel function for graphical inspection of the time series characteristics.

The specification of the test regression is determined by the functional argument compdet, which is a three element vector. The inclusion of an intercept, seasonal dummies, or a linear trend is chosen by providing a 1 at the corresponding place in the vector and a 0 else. The inclusion of lagged seasonal differences is set by the argument selecP, which can be a specific order or an automatic selection is done according to either the Akaike information criterion (AIC)– or Bayesian information criterion (BIC)–criterion, a Ljung–Box test, or only the significant lags are retained. Finally, additional dummy regressors can be included by the arguments Mvfic and VFEp. These arguments enable the researcher to model explicitly structural breaks in the seasonal means and increasing seasonal variation as was suggested as an amendment to the HEGY–test by Franses and Hobijn [30].

In Rcode example 5.3, the HEGY–test is applied to the logarithm of real disposable income in the United Kingdom from 1955:Q1 until 1984:Q4. This series is contained in the data set UKconinc in the contributed package urca and was used in Hylleberg et $al.$ [49]. The authors have chosen the lags 1, 4, 5 in the augmented test regression 5.18 and have run a combination of the deterministic regressors (see command lines 7 to 11). The t ratios of π_i for $i = 1, \ldots, 4$ can be retrieved by object[[1]]$tpi, and the F statistics of the Lagrange-multiplier test are stored in object[[1]]$Fpi. Finally, the significance of the deterministic regressors can be checked by inspecting object$compdeter.

Rcode 5.3 HEGY–test for seasonal unit roots

```
library(urca)                                                    1
library(uroot)                                                   2
data(UKconinc)                                                   3
attach(UKconinc)                                                 4
incl <- ts(incl, start=c(1955,1), end=c(1984,4), frequency       5
    =4)
incllist <- list(vari=incl, s=4, t0=c(1955, 1), N=length(        6
    incl))
HEGY000 <- HEGY.test(label=incllist, compdet=c(0, 0, 0),         7
    selecP=c(1,4,5), Mvfic=0, VFEp=0, showcat=FALSE)
HEGY100 <- HEGY.test(label=incllist, compdet=c(1, 0, 0),         8
    selecP=c(1,4,5), Mvfic=0, VFEp=0, showcat=FALSE)
HEGY110 <- HEGY.test(label=incllist, compdet=c(1, 1, 0),         9
    selecP=c(1,4,5), Mvfic=0, VFEp=0, showcat=FALSE)
HEGY101 <- HEGY.test(label=incllist, compdet=c(1, 0, 1),        10
    selecP=c(1,4,5), Mvfic=0, VFEp=0, showcat=FALSE)
HEGY111 <- HEGY.test(label=incllist, compdet=c(1, 1, 1),        11
    selecP=c(1,4,5), Mvfic=0, VFEp=0, showcat=FALSE)
```

The test results are provided in Table 5.5, where in the first column the specification is given as I for the inclusion of an intercept, SD for the inclusion of seasonal dummies and Tr abbreviates a linear trend. The reported t ratios for π_3 and π_4 as well as the F statistic $\pi_3 \cap \pi_4$ are all significant at the 5% level.

	t:π_1	t:π_2	t:π_3	t:π_4	F:$\pi_3 \cap \pi_4$
none	2.61	−1.44	−2.35	−2.51	5.68
I	−1.50	−1.46	−2.38	−2.51	5.75
I, SD	−1.56	−2.38	−4.19	−3.89	14.73
I, Tr	−2.73	−1.46	−2.52	−2.24	5.46
I, SD, TR	−2.48	−2.30	−4.28	−3.46	13.74

Table 5.5. HEGY–test: Real disposable income in the United Kingdom

Hence, the authors conclude that the null hypothesis of a unit root at the zero frequency cannot be rejected. However, the null hypothesis for the conjugate complex roots must be rejected.

Summary

In this chapter, the analysis of integrated time series has been amended in two important ways. First, it has been shown that in the case of a structural break the test conclusion about the presence of a unit root in a time series can be biased toward accepting it. Therefore, if *a-priori* knowledge of a structural shift is existent or by eye-spotting a break in the series is evident, one should either use the Perron or Zivot and Andrews test, respectively. Second, if the correlogram gives hindsight of seasonality in the time series, one should apply a seasonal unit root test. A complete analysis of a possibly integrated time series would therefore begin by testing wether breaks and/or stochastic seasonality is existent and dependent on this outcome unit root tests should be applied as shown in Chapter 4. After all, when the null hypothesis of a unit root must be rejected, it should be checked whether long memory behavior is present as shown in Section 2.3.

Exercises

1. Write a function that displays the critical values for models of type (A), (B), and (C) as in Perron [75].
2. Write a function that estimates the models of type (A), (B), and (C) as in Equations 5.5a through 5.5c.
3. Combine your functions from Excercise 1 and 2 so that now the functions return the relevant critical values for a prior specified significance level.
4. Write a function for the seasonal unit root test proposed by Osborn *et al.* [71].
5. Apply this function to the log of real disposable income in the United Kingdom as contained in the data set UKconinc and compare with the results reported in Table 5.5.

Part III

Cointegration

6

Single Equation Methods

This is the first chapter of the third and last part of this book. The cointegration methodology is first exemplified for the case of single equation models. The Engle–Granger two-step procedure is exemplified by estimating a consumption function and its error-correction form for the United Kingdom as in Holden and Perman [45]. In the ensuing section the method proposed by Phillips and Ouliaris [81] is applied to the same data set. The application and inferences of a vector error correction model (VECM) is saved for the second chapter.

6.1 Engle–Granger Two-Step Procedure

Recall from Section 3.2 that the first step of the Engle–Granger two-step procedure consists of estimating the long-run relationship as in Equation 3.4. Holden and Perman [45] applied this procedure to the estimation of a consumption function for the United Kingdom. The integration order of the consumption series has already been discussed in Section 4.1, and the determination of the integration order of the income and wealth series has been given as Exercise 1 in Chapter 4. In the following discussion, we will treat all series as $I(1)$ although the result for the wealth series is ambigous.[1] The authors regressed consumption on income and wealth for the sample period from 1967:Q2 until 1991:Q2. In Rcode example 6.1, the data set `Raotbl3` is loaded and the series are transformed into time series objects (see command lines 3 to 7). The selection of the sample period is easily achieved by the function `windows()` in command line 8. By slightly diverging from the analysis as in Holden and Perman, the long-run relationships for each of the series, *i.e.*, consumption, income, and wealth enter as endogenous variables separately, are simply estimated by ordinary least-squares (OLS) (see command lines 9 to 11).

---

[1] When a broken trend is allowed in the data-generated process for the wealth series, the authors concluded that the unit root hypothesis must be rejected on the basis of the test proposed by Perron [75]. This finding is confirmed by applying the Zivot and Andrews [107] test for a model with a constant, trend, and four lags (see Section 5.1 for a discussion of both tests).

Rcode 6.1 Engle–Granger: Long-run relationship of consumption, income, and wealth in the United Kingdom

```
library(urca)                                                              1
library(tseries)                                                           2
data(Raotbl3)                                                              3
attach(Raotbl3)                                                            4
lc <- ts(lc, start=c(1966,4), end=c(1991,2), frequency=4)                  5
li <- ts(li, start=c(1966,4), end=c(1991,2), frequency=4)                  6
lw <- ts(lw, start=c(1966,4), end=c(1991,2), frequency=4)                  7
ukcons <- window(cbind(lc, li, lw), start=c(1967, 2), end=c               8
    (1991,2))
lc.eq <- summary(lm(lc ~ li + lw, data=ukcons))                           9
li.eq <- summary(lm(li ~ lc + lw, data=ukcons))                          10
lw.eq <- summary(lm(lw ~ li + lc, data=ukcons))                          11
error.lc <- ts(resid(lc.eq), start=c(1967,2), end=c(1991,2)              12
    , frequency=4)
error.li <- ts(resid(li.eq), start=c(1967,2), end=c(1991,2)              13
    , frequency=4)
error.lw <- ts(resid(lw.eq), start=c(1967,2), end=c(1991,2)              14
    , frequency=4)
ci.lc <- ur.df(error.lc, lags=1, type='none')                           15
ci.li <- ur.df(error.li, lags=1, type='none')                           16
ci.lw <- ur.df(error.lw, lags=1, type='none')                           17
jb.lc <- jarque.bera.test(error.lc)                                     18
jb.li <- jarque.bera.test(error.li)                                     19
jb.lw <- jarque.bera.test(error.lw)                                     20
```

The residuals of these three long-run relationships are stored as objects `error.lc`, `error.li`, and `error.lw`, respectively. An augmented Dickey–Fuller (ADF)–type test is applied to the residuals of each equation for testing whether the variables are cointegrated or not (see command lines 15 to 17). Please note that one must now use the critical values found in MacKinnon [62] or Engle and Yoo [25].

	ADF	JB	JB p-value
consumption	−4.14	0.66	0.72
income	−4.06	0.07	0.97
wealth	−2.71	3.25	0.20

Table 6.1. Engle–Granger: Cointegration test

The test statistics imply cointegration for the consumption and income function that are significant at the 5% level, given a critical value of −3.83, but not for the wealth equation, thereby stressing the finding that this series should be considered as stationary with a broken trend. Furthermore, the Jarque–Bera test indicates that the null hypothesis of normality cannot be rejected for all equations. In the next step, the error correction models (ECMs) for the consumption and income functions are specified as in Equations 3.5a and 3.5b. In Rcode example 6.2, the necessary first differences of the series and its lagged values are created as well as the series for the lagged error term by one period.

Rcode 6.2 Engle–Granger: ECMs for consumption and income functions in the United Kingdom

---

```
ukcons2 <- ts(embed(diff(ukcons), dim=2), start=c(1967,4),     1
    freq=4)
colnames(ukcons2) <- c('lc.d', 'li.d', 'lw.d', 'lc.d1', 'li.   2
    d1', 'lw.d1')
error.ecm1 <- window(lag(error.lc, k=-1), start=c(1967,4),     3
    end=c(1991, 2))
error.ecm2 <- window(lag(error.li, k=-1), start=c(1967,4),     4
    end=c(1991, 2))
ecm.eq1 <- lm(lc.d ~ error.ecm1 + lc.d1 + li.d1 + lw.d1,       5
    data=ukcons2)
ecm.eq2 <- lm(li.d ~ error.ecm2 + lc.d1 + li.d1 + lw.d1,       6
    data=ukcons2)
```

The regression results for both ECMs are depicted in Tables 6.2 and 6.3. It should be restressed at this point, that if two series are cointegrated, then there should be Granger-causation in at least one direction. That is, at least one coefficient of the error term should enter the Equations 3.5a or 3.5b significantly and with the correct sign, *i.e.*, negative. Hence, even if the lagged differences of the income and consumption regressors do not enter significantly, the levels might through the residuals and hence Granger-cause consumption and/or income.

	Estimate	Std. Error	t value	$\Pr(>|t|)$
(Intercept)	0.0058	0.0015	3.86	0.0002
error.ecm1	0.0625	0.0984	0.64	0.5268
lc.d1	−0.2856	0.1158	−2.47	0.0156
li.d1	0.2614	0.0864	3.03	0.0032
lw.d1	0.0827	0.0317	2.61	0.0106

Table 6.2. Engle–Granger: ECM for the consumption function

The coefficient of the error correction term in the consumption function does not enter significantly and has the wrong sign (see Table 6.2). On the contrary, the error correction term does enter significantly and has the correct sign in the income equation (see Table 6.3). The error of the last period is worked off by one half, although the lagged differences of the remaining regressors do not enter significantly into the ECM. These results imply Granger-causation from consumption to income.

	Estimate	Std. Error	t value	Pr(>|t|)
(Intercept)	0.0066	0.0019	3.53	0.0006
error.ecm2	−0.5395	0.1142	−4.72	0.0000
lc.d1	−0.1496	0.1464	−1.02	0.3096
li.d1	−0.0060	0.1085	−0.06	0.9558
lw.d1	0.0627	0.0398	1.58	0.1187

Table 6.3. Engle–Granger: ECM for the income function

6.2 Phillips–Ouliaris Method

In Section 6.1 and in Section 3.2, it has been shown that the second step of the Engle–Granger method is an ADF–type test applied to the residuals of the long-run equation. Phillips and Ouliaris [81] introduced two residual-based tests, namely a variance ratio and a multivariate trace statistic. The latter of these tests has the advantage that it is invariant to normalization, $i.e.$, which variable is taken as endogenous. Both tests are based on the residuals of the first-order vector autoregression:

$$z_t = \hat{\boldsymbol{\Pi}} z_{t-1} + \hat{\boldsymbol{x}} i_t , \qquad (6.1)$$

where z_t is partioned as $z_t = (y_t, \boldsymbol{x}_t')$ with a dimension of $\boldsymbol{x}_t$ equal to ($m = n + 1$). The variance ratio statistic $\hat{P}_u$ is then defined as

$$\hat{P}_u = \frac{T \hat{\omega}_{11 \cdot 2}}{T^{-1} \sum_{t=1}^{T} \hat{u}_t^2} , \qquad (6.2)$$

where $\hat{u}_t$ are the residuals of the long-run equation $y_t = \hat{\boldsymbol{\beta}}' \boldsymbol{x}_t + \hat{u}_t$. The conditional covariance $\hat{\omega}_{11 \cdot 2}$ is derived from the covariance matrix $\hat{\Omega}$ of $\hat{\boldsymbol{\xi}}_t$, $i.e,$ the residuals of Equation 6.1, and is defined as

$$\hat{\omega}_{11 \cdot 2} = \hat{\omega}_{11} - \hat{\omega}_{21}' \hat{\boldsymbol{\Omega}}_{22}^{-1} \hat{\omega}_{21} , \qquad (6.3)$$

where the covariance matrix $\hat{\boldsymbol{\Omega}}$ has been partitioned as

$$\hat{\boldsymbol{\Omega}} = \begin{bmatrix} \hat{\omega}_{11} & \hat{\omega}_{21} \\ \hat{\omega}_{21} & \hat{\boldsymbol{\Omega}}_{22} \end{bmatrix}, \tag{6.4}$$

and is estimated as

$$\hat{\boldsymbol{\Omega}} = T^{-1} \sum_{t=1}^{T} \hat{\boldsymbol{\xi}}_t' \hat{\boldsymbol{\xi}}_t + T^{-1} \sum_{s=1}^{l} w_{sl} \sum_{t=1}^{T} (\hat{\boldsymbol{\xi}}_t \hat{\boldsymbol{\xi}}_{t-s}' + \hat{\boldsymbol{\xi}}_{t-s} \hat{\boldsymbol{\xi}}_t'), \tag{6.5}$$

with weighting function $w_{sl} = 1 - s/(l+1)$. Therefore, the variance ratio statistic measures the size of the residual variance from the cointegrating regression of y_t on $\boldsymbol{x}_t$ against that of the conditional variance of y_t given $\boldsymbol{x}_t$. In the case of cointegration, the test statistic should stabilize to a constant, whereas if a spurious relationship is present, this would be reflected in a divergent variance of the long-run equation residuals from the conditional variance. Critical values of this test statistic have been tabulated in Phillips and Ouliaris [81].

The multivariate trace statistic, denoted as $\hat{P}_z$, is defined as

$$\hat{P}_z = T tr(\hat{\boldsymbol{\Omega}} \boldsymbol{M}_{zz}^{-1}), \tag{6.6}$$

with $\boldsymbol{M}_{zz} = t^{-1} \sum_{t=1}^{T} \boldsymbol{z}_t \boldsymbol{z}_t'$ and $\hat{\boldsymbol{\Omega}}$ estimated as in equation 6.5. Critical values for this test statististic are provided in Phillips and Ouliaris too.

Both tests are implemented in the function `ca.po()` in the contributed package `urca`. Besides the specification of the test type, the inclusion of deterministic regressors can be set *via* the argument `demean` and the lag length for estimating the long-run variance-covariance matrix $\hat{\boldsymbol{\Omega}}$ can be set with the argument `lag`. Because a matrix inversion is needed in the calculation of the test statistics, one can pass a tolerance level to the implicitly used function `solve()` *via* the argument `tol`. The default value is `NULL`.

In Rcode example 6.3, both test types are applied to the same data set as before. The results are provided in Table 6.4. The variance ratio test statistic does indicate a spurious relationship. It should be of no suprise, because the first column of the data set `ukcons` is the consumption series. Therefore, the test conclusion is the same as using the two-step Engle–Granger procedure.

Rcode 6.3 Phillips–Ouliaris: Long-run relationship of consumption, income, and wealth in the United Kingdom

---

```
library(urca)                                                          1
data(Raotbl3)                                                          2
attach(Raotbl3)                                                        3
lc <- ts(lc, start=c(1966,4), end=c(1991,2), frequency=4)             4
li <- ts(li, start=c(1966,4), end=c(1991,2), frequency=4)             5
lw <- ts(lw, start=c(1966,4), end=c(1991,2), frequency=4)             6
ukcons <- window(cbind(lc, li, lw), start=c(1967, 2), end=c           7
    (1991,2))
pu.test <- summary(ca.po(ukcons, demean='const', type='Pu'))         8
pz.test <- summary(ca.po(ukcons, demean='const', type='Pz'))         9
```

---

	test statistic	10%	5%	1%
$\hat{P}_u$	58.91	80.20	89.76	109.45
$\hat{P}_z$	88.03	33.70	40.53	53.87

Table 6.4. Phillips–Ouliaris: Cointegration test

However, matters are different if one uses the $\hat{P}_z$ statistic. From the inclusion of the wealth series, which is considered as stationary around a broken trend line, the statistic does indicate a spurious relationship. Please note that normalization of the long-run equation does not affect this test statistic.

By now we have only discussed single equation methods and how such methods can be fairly easily applied in R. One deficiency of these methods is that one can only estimate a single cointegration relationship. However, if one deals with more than two time series, it is possible that more than only one cointegrating relationship exists as has been pointed out in Section 3.3. The estimation and inference of VECMs are the subject of the next chapter.

Summary

In the first chapter of Part III, two single equation methods have been presented. The advantage of the Engle–Granger two-step procedure is its ease of implementation. However, the results are dependent on how the long-run equation is specified. In most cases, it might be obvious which variable enters on the left-hand side of the equation, *i.e.*, to which variable the cointegrating vector should be normalized. Unfortunately, this is only true in most cases, and as anecdotal evidence, an income function rather than a consumption function could have been specified as an ECM in Rcode example 6.2. It is therefore advisable to employ the cointegration test of Phillips and Ouliaris that is irrelevant to normalization.

As mentioned, the insights gained with respect to the cointegrating relationship are limited in the case of more than two variables. The topic of the next chapter is therefore dedicated to the inference in cointegrated systems.

Exercises

1. Consider the data sets `Raotbl1` and `Raotbl1` in the contributed package `urca`. Your goal is to specify ECMs for real money demand functions by using different monetary aggregates.
   a) Determine the integration order of the series first.
   b) Estimate the long-run equations.
   c) Can you find cointegration relations for the different money demand functions?
   d) Specify the ECMs and interpret your results with respect to the error correcting term.
2. Consider the data set `Raotbl6` in the contributed package `urca`. Specify a Phillips-Curve model in error correction form as in Mehra [68].
   a) Determine the integration order of the price level, unit labor cost, and output gap variable first.
   b) Estimate the long-run equation and test for cointegration. Hereby, employ the Phillips–Ouliaris tests too.
   c) Specify an ECM, and discuss your findings.

Multiple Equation Methods

In this chapter, the powerful tool of likelihood-based inference in cointegrated vector autoregressive models (VECMs) is discussed. In a first step, the specification and assumptions of a VECM are introduced. In the following sections, the problems of determing the cointegration rank, the testing for weak exogenity, as well as the testing of various restrictions placed on the cointegrating vectors are discussed. Finally, the topic of VECMs that are contaminated by a one-time structural shift is presented and how this kind of models can be estimated.

7.1 The Vector Error Correction Model (VECM)

7.1.1 Specification and Assumptions

In this section, the results in Johansen and Juselius [55] are replicated. In this article, the authors are testing structural hypotheses in a multivariate cointegration context of the *purchasing power parity* and the *uncovered interest parity* for the United Kingdom. They use quarterly data spanning a range from 1972:Q1 to 1987:Q2. As variables p_1: the U.K. wholesale price index, p_2: the trade-weighted foreign wholesale price index, e_{12}: the U.K. effective exchange rate, i_1: the three-month treasury bill rate in the United Kingdom and i_2: the three-month Eurodollar interest rate enter into the VECM. To cope with the oil crises at the beginning of the sample period, the world oil price (contemporanesously and lagged once), denoted as $doilp_0$ and $doilp_1$, respectively, is included as an exogenous regressor too. These variables, expressed in natural logarithms, are included in the data set UKpppuip contained in the package urca and depicted in Figure 7.1.

As a preliminary model, they settled on the following specification:

$$\boldsymbol{y}_t = \boldsymbol{\Gamma}_1 \Delta \boldsymbol{y}_{t-1} + \boldsymbol{c}_0 \Delta x_t + \boldsymbol{c}_1 \Delta x_{t-1} + \boldsymbol{\Pi} \boldsymbol{y}_{t-2} + \boldsymbol{\mu} + \boldsymbol{\Phi} \boldsymbol{D}_t + \boldsymbol{\varepsilon}_t , \qquad (7.1)$$

where the vector $\boldsymbol{y}_t$ contains as elements $(p_1, p_2, e_{12}, i_1, i_2)'$ and x_t assigns the model exogenous world oil price $doilp_0$. In the matrix $\boldsymbol{D}_t$, centered seasonal dummy variables are included and the vector $\boldsymbol{\mu}$ is a vector of constants. The five-dimensional error process $\boldsymbol{\varepsilon}_t$ is assumed to be i.i.d. as $\mathcal{N}(\boldsymbol{0}, \boldsymbol{\Sigma})$ for $t = 1, \ldots, T$. This specification is the long-run form of a VECM (see Equation 3.8).

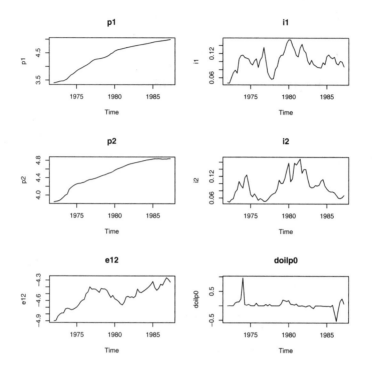

Fig. 7.1. Graphical display of purchasing power parity and uncovered interest rate parity for the United Kingdom

7.1.2 Determining the Cointegration Rank

The authors start by determining the cointegration rank. Because inferences on the cointegration space spanned by its vectors is dependent on the existence of linear trends in the data or not, they argued by ocular econometrics and logical reasoning that the price series have a linear trend that is consistent with the steady-state assumption of constant nominal price growth as implied by the economic theory, and therefore, the vector μ can be estimated without imposing any restrictions.[1]

---

[1] In the case of linear trends, the constant vector μ can be partioned as $\mu = \alpha\beta_0 + \alpha_\perp\gamma$, where β_0 is a $r \times 1$ vector of intercepts in the cointegration relations, $\alpha_\perp$ is a $n \times (n-r)$ matrix of full rank perpendicular to the columns of α, and γ is a $(n-r) \times 1$ vector of linear trend slopes. Therefore, the alternative hypothesis is $\alpha_\perp\gamma = 0$ and can be tested as shown in Johansen and Juselius [54]. This linear trend test is implemented in the package urca as lttest().

In Rcode example 7.1, the hypothesis $\mathcal{H}_1(r) : \boldsymbol{\Pi} = \boldsymbol{\alpha\beta'}$, *i.e.*, that $\boldsymbol{\Pi}$ is of reduced rank, is tested with the trace and the maximal eigenvalue statistic (see Equations 3.11 and 3.14).[2]

Before the results of these tests are discussed, the arguments of the function ca.jo() should be presented briefly. The data set is provided by x, and the test type is either **eigen** or **trace** for the maximal eigenvalue statistic or the trace statistic, respectively, where the default is the former. Whether a constant should be included in the cointegration relations can be set by the logical variable **constant**. The decision of whether the long-run or transitory form of the VECM should be estimated is determined by the argument **spec**. The default is **longrun**. The inclusion of centered seasonal dummy variables can be set by providing the corresponding seasonality as integer, *e.g.*, 4 for quarterly data. Model exogenous regressors can be provided by setting **dumvar** accordingly. Finally, the appropriate critical values are chosen to be either **A1**, **A2**, or **A3** (see the appendix in Johansen and Juselius [54]). The table **A1** refers to the case of $\boldsymbol{\alpha}_{\perp}\boldsymbol{\mu} \neq 0$, the table **A2** is relevant for the case of $\boldsymbol{\alpha}_{\perp}\boldsymbol{\mu} = 0$, and the table **A3** is used if $\boldsymbol{\mu} = \boldsymbol{\alpha\beta}'_0$.

Rcode 7.1 Johansen–Juselius: Unrestricted cointegration

```
library (urca)                                                    1
data (UKpppuip)                                                   2
names (UKpppuip)                                                  3
attach (UKpppuip)                                                 4
dat1 <- cbind (p1, p2, e12, i1, i2)                              5
dat2 <- cbind (doilp0, doilp1)                                   6
args ('ca.jo')                                                   7
H1 <- ca.jo (dat1, type='trace', K=2, season=4, dumvar=dat2,    8
   ctable='A1')
H1.trace <- summary (ca.jo (dat1, type='trace', K=2, season     9
   =4, dumvar=dat2, ctable='A1'))
H1.eigen <- summary (ca.jo (dat1, type='eigen', K=2, season    10
   =4, dumvar=dat2, ctable='A1'))
```

In Tables 7.1 and 7.2, the results of the two tests are given. If considering the maximal eigenvalue statistic, the hypothesis of no cointegration cannot

---

[2] Incidentally, the internal examples of the functions for estimating and testing VECMs in the package **urca** are a replication of Johansen and Juselius [54], *i.e.*, the analysis of money demand functions for Denmark and Finland. For example, by typing example(ca.jo()), the results of determining the cointegration rank in this study are displayed. The reader is encouraged to work through these examples to foster understanding and comprehension of the method and the tools available. It is of course best accomplished by having a copy of the above-cited article at hand.

be rejected at the 5% level. However, the trace statistic indicates a cointegra-

	test statistic	10%	5%	1%
r <= 4	5.19	2.82	3.96	6.94
r <= 3	6.48	12.10	14.04	17.94
r <= 2	17.59	18.70	20.78	25.52
r <= 1	20.16	24.71	27.17	31.94
r = 0	31.33	30.77	33.18	38.34

Table 7.1. Cointegration rank: Maximal eigenvalue statistic

	test statistic	10%	5%	1%
r <= 4	5.19	2.82	3.96	6.94
r <= 3	11.67	13.34	15.20	19.31
r <= 2	29.26	26.79	29.51	35.40
r <= 1	49.42	43.96	47.18	53.79
r = 0	80.75	65.06	68.91	76.95

Table 7.2. Cointegration rank: Trace statistic

tion space of $r = 2$, given a 5% significance level. Hence, the two tests are yielding contradictary conclusions about the cointegration rank. The decision about the cointegration rank is even more complicated by the fact that the estimated second and third eigenvalue are approximately equal (0.407, 0.285, 0.254, 0.102, 0.083) and therefore suggest a third stationary linear combination. The eigenvalues are in the slot `lambda` of an object adhering to class `ca.jo`. To settle for a working assumption about the cointegration rank, the authors investigated the $\hat{\beta}$ and $\hat{\alpha}$ matrices as well as the estimated cointegration relations $\hat{\beta}'_i X_t$ and the ones that are corrected for short-term influences, $\hat{\beta}'_i R_{Kt}$. The $\hat{\beta}$ and $\hat{\alpha}$ matrices are stored in the slots V and W, respectively, and the matrix R_K in the slot RK of a class `ca.jo` object. For ease of comparison with table 3 in Johansen and Juselius [55], the elements in the cointegration and loadings matrix have been normalized accordingly and are displayed in Tables 7.3 and 7.4. In Rcode example 7.2, the commands for calculating the above-mentioned figures are displayed. The authors argued that the values of $\hat{\alpha}_{i.2}$ for $i = 1, 2, 3$ are close to zero for the second cointegration vector, and therefore, the low estimated value of the second eigenvalue $\hat{\lambda}_2$ can be attributed to this fact. Furthermore, the power of the test is low in cases when the cointegration relation is close to the nonstationary boundary. This artifact can be the result of a slow speed of adjustment as is often the case in empirical work because of transaction costs and other obstacles that place a hindrance to a quick equilibrium adjustment. Johansen and Juselius investigated visually

Rcode 7.2 $\mathcal{H}_1$ model: Transformations and cointegration relations

```
beta <- H1@V                                                              1
beta[,2] <- beta[,2]/beta[4,2]                                           2
beta[,3] <- beta[,3]/beta[4,3]                                           3
alpha <- H1@PI%*%solve(t(beta))                                          4
beta1 <- cbind(beta[,1:2], H1@V[,3:5])                                   5
ci.1 <- ts((H1@x%*%beta1)[-c(1,2),], start=c(1972, 3), end=c            6
    (1987, 2), frequency=4)
ci.2 <- ts(H1@RK%*%beta1, start=c(1972, 3), end=c(1987, 2),             7
    frequency=4)
```

	$\hat{\beta}_{0.1}$	$\hat{\beta}_{0.2}$	$\hat{v}_3$	$\hat{v}_4$	$\hat{v}_5$
p1	1.00	0.03	0.36	1.00	1.00
p2	−0.91	−0.03	−0.46	−2.40	−1.45
e12	−0.93	−0.10	0.41	1.12	−0.48
i1	−3.37	1.00	1.00	−0.41	2.28
i2	−1.89	−0.93	−1.03	2.99	0.76

Table 7.3. $\mathcal{H}_1$ model: Eigenvectors

	$\hat{\alpha}_{0.1}$	$\hat{\alpha}_{0.2}$	$\hat{w}_3$	$\hat{w}_4$	$\hat{w}_5$
1	−0.07	0.04	−0.01	0.00	−0.01
2	−0.02	0.00	−0.04	0.01	0.01
3	0.10	−0.01	−0.15	−0.04	−0.05
4	0.03	−0.15	−0.03	0.01	−0.02
5	0.06	0.29	0.01	0.03	−0.01

Table 7.4. $\mathcal{H}_1$ model: Weights

the cointegration relationships. The first two of them are depicted in Figure 7.2. In the case of $r = 2$, the first two cointegration relationships should behave like stationary processes. However, because of short-run influences that overlay the adjustment process, this picture can be camouflaged. Hence, the authors also analyze the adjustment paths $\hat{\beta}'_i R_{kt}$ that take the short-run dynamics into account. Based on the test results, the elements in the $\hat{\alpha}$ matrix and the shape of the cointegration relation paths, Johansen and Juselius decided to stick with a cointegration order of $r = 2$.

Finally, in Table 7.5, the estimated $\hat{\Pi}$ is displayed. This matrix measures the combined effects of the two cointegration relations. The purchasing power parity hypothesis postulates a relationship between the two price indices and the exchange rate of $(\alpha_i, -\alpha_i, -\alpha_i)$. These relations are closely fullfilled for the first, third, and fourth equation.

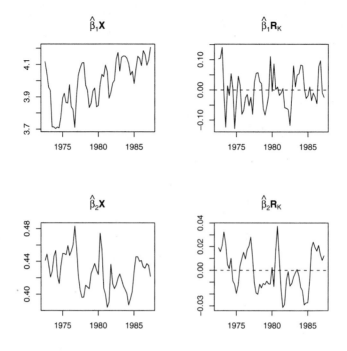

Fig. 7.2. Graphical display of the first two cointegration relations

	p1	p2	e12	i1	i2
1	−0.067	0.061	0.060	0.272	0.090
2	−0.018	0.016	0.016	0.064	0.030
3	0.101	−0.091	−0.093	−0.345	−0.186
4	0.030	−0.026	−0.018	−0.263	0.072
5	0.066	−0.062	−0.082	0.097	−0.382

Table 7.5. $\mathcal{H}_1(2)$ model: Coefficient matrix of the lagged variables in levels, $\hat{\Pi}$

7.1.3 Testing for Weak Exogenity

By now it has been assumed that all components of y_t are treated as endogenous variables. But sometimes we are interested in treating some of these components as exogenous, such that model simulations can be carried out by providing alternative paths for the exogenously assumed variables. Hence, a test is required for the full system if the hypothesis that some components of y_t are exogenous is valid. The test idea is to see whether zero retsrictions on the relevant rows of α hold. The restrictions are contained in the $(n \times m)$ matrix A such that $\mathcal{H}_4 : \alpha = A\Psi$, whereby the elements of the matrix Ψ contain the new unrestricted loadings. Johansen and Juselius [54] showed that this test can be traced down to an eigenvalue problem likewise to the one given in Equation 3.12. For calculating the new sets of residuals that enter the conditional concentrated likelihood function, it is convenient to introduce the $(N \times (N - m))$ matrix B that is orthogonal to A such that $B'A = 0$ or equivently $B'\alpha = 0$. Johansen [53] and Johansen and Juselius [54] then showed that the sets of residuals that enter the concentrated likelihood function are defined as

$$\tilde{R}_{at} = A'R_{0t} - \hat{S}_{ab}\hat{S}_{bb}^{-1}B'R_{0t} , \qquad (7.2a)$$

$$\tilde{R}_{kt} = R_{kt} - \hat{S}_{kb}\hat{S}_{bb}^{-1}B'R_{0t} , \qquad (7.2b)$$

$$\qquad (7.2c)$$

where R_{0t} and R_{kt} and the product moment matrices $\hat{S}_{ij}$ with $i,j = 0, k$ are given in Equation 3.10 and the product moment matrices $\hat{S}_{ab} =$, $\hat{S}_{ab} =$ and $\hat{S}_{kb}$ are defined as $\hat{S}_{ab} = A'\hat{S}_{00}B$, $\hat{S}_{bb} = B'\hat{S}_{00}B$, and $\hat{S}_{kb} = \hat{S}_{0k}B$. The relevant product moment matrices that enter the eigenvalue equation are defined as

$$\hat{S}_{ij.b} = \frac{1}{T}\sum_{t=1}^{T} \tilde{R}_{it}\tilde{R}'_{it} \text{ with } i,j = a,k . \qquad (7.3)$$

The maximum likelihood estimator for β under the hypothesis $\mathcal{H}_4 : \alpha = A\Psi$ is defined as the solution to the eigenvalue equation:

$$|\lambda\hat{S}_{kk.b} - \hat{S}_{ka.b}\hat{S}_{aa.b}^{-1}\hat{S}_{ak.b}| = 0 , \qquad (7.4)$$

with eigenvalues $\tilde{\lambda}_{4.1} > \tilde{\lambda}_{4.2} > \ldots > \tilde{\lambda}_{4.m} > \tilde{\lambda}_{4.m+1} = \ldots = \tilde{\lambda}_{4.p} = 0$ and the corresponding eigenvectors $\tilde{V}_4 = (\tilde{v}_{4.1}, \ldots, \tilde{v}_{4.p})$ are normalized such that $\tilde{V}'_4\hat{S}_{kk.b}\tilde{V}_4 = I$. The weighting matrix $\hat{\Psi}$ is given by

$$\hat{\Psi} = (A'A)^{-1}\hat{S}_{ak.b}\tilde{\beta}_4 \qquad (7.5)$$

with $\tilde{\beta}_4 = (\tilde{v}_{4.1}, \ldots, \tilde{v}_{4.r})$ under the maintained hypothesis of $\mathcal{H}_1(r)$. Johansen [52] proposed the following likelihood ratio statistic for testing the validity of $\mathcal{H}_4 : \alpha = A\Psi$ given $\mathcal{H}_1(r)$:

$$-2\ln(Q; \mathcal{H}_4|\mathcal{H}_1(r)) = T\sum_{i=1}^{r} \ln\{\frac{(1 - \tilde{\lambda}_{4.i})}{(1 - \hat{\lambda}_i)}\} , \qquad (7.6)$$

which is asymptotically distributed as χ^2 with $(r(N - m)$ degrees of freedom. This test statistic is implemented as function `alrtest()` in the package `urca`. Therefore, if $\mathcal{H}_4$ cannot be rejected, the VECM can be reduced to an m-dimensional system by conditioning to Δy_{2t}, where y_{2t} contain the exogenous variables.

The authors applied this test to the trade-weighted foreign wholesale price index and the three-month Eurodollar interest rate, respectively, $i.e.$, p2 and i2. The restriction matrices A_1 and A_2 are then given by

$$A_1 = \begin{bmatrix} 1 & 0 & 0 & 0 \\ 0 & 0 & 0 & 0 \\ 0 & 1 & 0 & 0 \\ 0 & 0 & 1 & 0 \\ 0 & 0 & 0 & 1 \end{bmatrix}, \quad A_2 = \begin{bmatrix} 1 & 0 & 0 & 0 \\ 0 & 1 & 0 & 0 \\ 0 & 0 & 1 & 0 \\ 0 & 0 & 0 & 1 \\ 0 & 0 & 0 & 0 \end{bmatrix}.$$

In Rcode example 7.3, these two test statistics are calculated. The matrices A_1 and A_2 are easily set up with the matrix() function. The ca.jo object corresponding to the model $\mathcal{H}_1(r = 2)$ has been created in Rcode example 7.1 as H1.

Rcode 7.3 $\mathcal{H}_4$ model: Testing for weak exogenity

```
A1 <- matrix(c(1,0,0,0,0, 0,0,1,0,0, 0,0,0,1,0, 0,0,0,0,1),      1
    nrow=5, ncol=4)
A2 <- matrix(c(1,0,0,0,0, 0,1,0,0,0, 0,0,1,0,0, 0,0,0,1,0),      2
    nrow=5, ncol=4)
H41 <- summary(alrtest(z=H1, A=A1, r=2))                         3
H42 <- summary(alrtest(z=H1, A=A2, r=2))                         4
```

The value of the test statistic is stored in the slot teststat and its marginal level of significance, the p-value, in the slot pval. The restricted eigenvalues can be retrieved by object@lambda and the associated eigenvectors, $i.e.$, the cointegration relations, by object@V. The new loadings as calculated as in Equation 7.5 are contained in the slot W. The results of the two tests are reported in Table 7.6. For the variable p2, the null hypothesis can-

	test statistic	p-value	$\tilde{\lambda}_1$	$\tilde{\lambda}_2$	$\tilde{\lambda}_3$	$\tilde{\lambda}_4$
$\mathcal{H}_{4.1}\|\mathcal{H}_1(r = 2)$	0.657	0.720	0.400	0.285	0.167	0.088
$\mathcal{H}_{4.2}\|\mathcal{H}_1(r = 2)$	4.384	0.112	0.387	0.256	0.194	0.086

Table 7.6. $\mathcal{H}_4$ model: Testing for weak exogenity

not be rejected, whereas for the interest rate variable, the result is borderline given a significance level of 10%.[3]

---

[3] Please note that the results differ from the ones reported in Johansen and Juselius [55]. The authors report slightly smaller values for the second eigenvalues for each test statistic. Qualitatively, the test conclusion is thereby only affected for

7.1.4 Testing Restrictions on $\boldsymbol{\beta}$

In this section, three statistical tests for validating different forms of restrictions on the $\boldsymbol{\beta}$ matrix are discussed. The hypotheses formulated about this matrix are, likewise to the test for the $\boldsymbol{\alpha}$ matrix, linear. Furthermore, these tests do not depend on the normalization of the cointegration relations. The first test can be used to test the validity of restrictions for all cointegration relations and is termed $\mathcal{H}_3$. This test was introduced in Johansen [51] and applied in Johansen and Juselius [54]. A theoretical exposition can also be found in Johansen [52][53]. In the second test, it is assumed that some r_1 of the r cointegration relations are assumed to be known and the remaining r_2 cointegration relations have to be estimated. This hypothesis is termed $\mathcal{H}_5$. Finally, in the last hypothesis $\mathcal{H}_6$, some restrictions are placed on the r_1 cointegration relations and the remaining r_2 ones are estimated without constraints. The last two restrictions have been introduced by the authors in Johansen and Juselius [55]. To summarize, these three hypotheses are listed below with the dimensions of the restriction matrices and the spaces of the cointegration relations to be tested:

i) $\mathcal{H}_3 : \boldsymbol{\beta} = \boldsymbol{H}_3\boldsymbol{\varphi}$ with $\boldsymbol{H}_3(N \times s)$, $\boldsymbol{\varphi}(s \times r)$ and $r \leq s \leq N$:
   $sp(\boldsymbol{\beta}) \subset sp(\boldsymbol{H}_3)$.
ii) $\mathcal{H}_5 : \boldsymbol{\beta} = (\boldsymbol{H}_5, \boldsymbol{\Psi}$ with $\boldsymbol{H}_5(N \times r_1)$, $\boldsymbol{\Psi}(N \times r_2)$, $r = r_1 + r_2$:
   $sp(\boldsymbol{H}_5) \subset sp(\boldsymbol{\beta})$.
iii) $\mathcal{H}_6 : \boldsymbol{\beta} = (\boldsymbol{H}_6\boldsymbol{\varphi}, \boldsymbol{\Psi}$ with $\boldsymbol{H}_6(N \times s)$, $\boldsymbol{\varphi}(s \times r_1)$, $\boldsymbol{\Psi}(N \times r_2)$, $r_1 \leq s \leq N$,
   $r = r_1 + r_2$:
   $\dim(sp(\boldsymbol{\beta}) \cap sp(\boldsymbol{H}_6)) \geq r_1$.

First, the hypothesis $\mathcal{H}_3$ is presented. Johansen showed that the estimator $\hat{\boldsymbol{\varphi}}$ under this hypothesis is the eigenvectors of

$$|\lambda \boldsymbol{H}_3' \hat{\boldsymbol{S}}_{kk} \boldsymbol{H}_3 - \boldsymbol{H}_3' \hat{\boldsymbol{S}}_{k0} \hat{\boldsymbol{S}}_{00}^{-1} \hat{\boldsymbol{S}}_{0k} \boldsymbol{H}_3| . \tag{7.7}$$

The solution to this equation gives the eigenvalues $\tilde{\lambda}_1 > \ldots > \tilde{\lambda}_s > 0$. The corresponding eigenvectors are denoted as $\tilde{\boldsymbol{V}} = (\tilde{\boldsymbol{v}}_1, \ldots, \tilde{\boldsymbol{v}}_s)$. The estimate of $\hat{\boldsymbol{\varphi}}$ is then given as $(\tilde{\boldsymbol{v}}_1, \ldots, \tilde{\boldsymbol{v}}_r)$, and therefore, $\hat{\boldsymbol{\beta}} = \boldsymbol{H}_3(\tilde{\boldsymbol{v}}_1, \ldots, \tilde{\boldsymbol{v}}_r)$.

The hypothesis $\mathcal{H}_3$ given $\mathcal{H}_1(r)$ can be tested with a likelihood ratio test defined as

$$-2\ln(Q; \mathcal{H}_3|\mathcal{H}_1(r)) = T \sum_{i=1}^{r} ln\left\{ \frac{(1 - \tilde{\lambda}_{3.i})}{(1 - \hat{\lambda}_i)} \right\} , \tag{7.8}$$

which is asymptotically distributed as χ^2 with $r(N - s)$ degrees of freedom. This test is implemented as function blrtest() in the contributed package

the second hypothesis where the authors report a value of 6.34, which clearly indicates that the three-month Eurordollar interest rate cannot be considered as weakly exogenous for β.

urca.

Johansen and Juselius used this test statistic for validating the purchasing power parity and for testing whether the interest rate differential enters all cointegration relations. The purchasing power parity states that the variables p_1, p_2, and e_{12} enter the cointegration relations proportionally as $(1, -1, -1)$; *i.e.*, if a stationary combination of these variables exists, then it must enter the cointegration relations as $(a_i, -a_i, -a_i, *, *)$ with $i = 1, \ldots, r$. Likewise, the restriction to test whether the interest rate differential, *i.e.*, $i_1 - i_2$, enters all cointegration relations is given by the proportional cointegration vector $(1, -1)$, and hence, this can be formulated as $(*, *, *, b_i, -b_i)$ for $i = 1, \ldots, r$. Therefore, the two restrictions matrices $\boldsymbol{H}_{3.1}$ and $\boldsymbol{H}_{3.2}$ are written as

$$\boldsymbol{H}_{3.1} = \begin{bmatrix} 1 & 0 & 0 \\ -1 & 0 & 0 \\ -1 & 0 & 0 \\ 0 & 1 & 0 \\ 0 & 0 & 1 \end{bmatrix} \quad , \quad \boldsymbol{H}_{3.2} = \begin{bmatrix} 1 & 0 & 0 & 0 \\ 0 & 1 & 0 & 0 \\ 0 & 0 & 1 & 0 \\ 0 & 0 & 0 & 1 \\ 0 & 0 & 0 & -1 \end{bmatrix}$$

The Rcode for conducting these two tests is displayed in Rcode example 7.4. As with Rcode example 7.3, an unrestricted model $\mathcal{H}_1$ the object H1 from Rcode example 7.1 has been used.

Rcode 7.4 $\mathcal{H}_3$ model: Testing for restrictions in all cointegration relations

```
H.31 <- matrix(c(1,-1,-1,0,0, 0,0,0,1,0, 0,0,0,0,1), c(5,3))    1
H.32 <- matrix(c                                                 2
    (1,0,0,0,0, 0,1,0,0,0, 0,0,1,0,0, 0,0,0,1,-1), c(5,4))
H31 <- summary(blrtest(z=H1, H=H.31, r=2))                       3
H32 <- summary(blrtest(z=H1, H=H.32, r=2))                       4
```

---

	test statistic	p-value	λ_1	λ_2	λ_3	λ_4
$\mathcal{H}_{3.1}\|\mathcal{H}_1(r=2)$	2.761	0.599	0.386	0.278	0.090	
$\mathcal{H}_{3.2}\|\mathcal{H}_1(r=2)$	13.709	0.001	0.286	0.254	0.146	0.093

Table 7.7. $\mathcal{H}_3$ model: Restriction in all cointegration relations

The results of the test are displayed in Table 7.7. In the case of the purchasing power parity, the model hypothesis $\mathcal{H}_3$ cannot be rejected. This result mirrors the closely found $(a_i, -a_i, -a_i, *, *)$ relations in the estimated $\hat{\boldsymbol{\Pi}}$ matrix for two cointegration relations as provided in Table 7.5.[4]

Next, the authors tested directly whether $(1, -1, -1, 0, 0)' \boldsymbol{y}_t$ and $(0, 0, 0, 1, -1)' \boldsymbol{y}_t$ each constitute a stationary relation. Hypotheses of such type can be tested with the model $\mathcal{H}_5$, in which some cointegration relations are assumed to be known.

To test the hypothesis $\mathcal{H}_5$, the partial weighting matrix corresponding to $\boldsymbol{H}_5$ is concentrated out of the likelihood function. It is achieved by regressing $\boldsymbol{R}_{0t}$ and $\boldsymbol{R}_{kt}$ on $\boldsymbol{H}_5' \boldsymbol{R}_{kt}$ and thereby obtaining the new sets of residuals:

$$\boldsymbol{R}_{0.ht} = \boldsymbol{R}_{0t} - \hat{\boldsymbol{S}}_{0k} \boldsymbol{H}_5 (\boldsymbol{H}_5' \hat{\boldsymbol{S}}_{kk} \boldsymbol{H}_5)^{-1} \boldsymbol{H}_5 \boldsymbol{R}_{kt} , \qquad (7.9a)$$

$$\boldsymbol{R}_{k.ht} = \boldsymbol{R}_{kt} - \hat{\boldsymbol{S}}_{kk} \boldsymbol{H}_5 (\boldsymbol{H}_5' \hat{\boldsymbol{S}}_{kk} \boldsymbol{H}_5)^{-1} \boldsymbol{H}_5 \boldsymbol{R}_{kt} . \qquad (7.9b)$$

The new product moment matrices are then calculated as

$$\hat{\boldsymbol{S}}_{ij.h} = \hat{\boldsymbol{S}}_{ij} - \hat{\boldsymbol{S}}_{ik} \boldsymbol{H}_5 (\boldsymbol{H}_5' \hat{\boldsymbol{S}}_{kk} \boldsymbol{H})^{-1} \boldsymbol{H}' \hat{\boldsymbol{S}}_{kj} \text{ for } i, j = 0, k . \qquad (7.10)$$

An estimate of the partially unknown cointegration relations $\hat{\boldsymbol{\Psi}}$ is obtained by solving two eigenvalue problems. First, step the $N - r_1$ eigenvalues of

$$|\tau \boldsymbol{I} - \hat{\boldsymbol{S}}_{kk.h}| = 0 \qquad (7.11)$$

are retrieved and the auxilliary matrix $\boldsymbol{C}$ is calculated as

$$\boldsymbol{C} = (\boldsymbol{e}_1, \boldsymbol{e}_2, \dots, \boldsymbol{e}_{N-r_1}) \begin{pmatrix} \tau_1^{-1/2} & 0 & \cdots & \cdots & 0 \\ 0 & \tau_2^{-1/2} & 0 & \cdots & 0 \\ \vdots & 0 & \ddots & & \vdots \\ \vdots & \vdots & & \ddots & 0 \\ 0 & 0 & \cdots & 0 & \tau_{N-r_1}^{-1/2} \end{pmatrix} , \qquad (7.12)$$

where $(\boldsymbol{e}_1, \boldsymbol{e}_2, \dots, \boldsymbol{e}_{N-r_1})$ are the eigenvectors belonging to $\boldsymbol{\tau}$. The matrix $\boldsymbol{C}$ then enters the second eigenvalue problem:

$$|\lambda \boldsymbol{I} - \boldsymbol{C}' \hat{\boldsymbol{S}}_{k0.h} \hat{\boldsymbol{S}}_{00.h}^{-1} \hat{bm} \boldsymbol{S}_{0k.h} \boldsymbol{C}| = 0 , \qquad (7.13)$$

which yields the eigenvalues $\tilde{\lambda}_1 > \dots > \tilde{\lambda}_{N-r_1} > 0$ and eigenvectors $\tilde{\boldsymbol{V}} = (\tilde{\boldsymbol{v}}_1, \dots, \tilde{\boldsymbol{v}}_{N-r_1})$. The partial cointegration relations are then estimated

---

[4] Incidentally, the two model hypotheses $\mathcal{H}_4$ and $\mathcal{H}_3$ for a given unrestricted model $\mathcal{H}_1(r)$ can be tested jointly, as shown in Johansen and Juselius [54], for instance. This combined test is implemented as function `ablrtest()` in the contributed package `urca`. An example of its application is provided in the citation above, which is mirrored in `example(ablrtest)`.

as $\hat{\boldsymbol{\Psi}} = \boldsymbol{C}(\tilde{\boldsymbol{v}}_1, \ldots, \tilde{\boldsymbol{v}}_{r_2})$, and therefore, the cointegration relations are given as $\hat{\boldsymbol{\beta}} = (\boldsymbol{H}_5, \hat{\boldsymbol{\Psi}})$.

Finally, for calculating the likehood ratio test statistic, the eigenvalues $\boldsymbol{\rho}$ have to be extracted from

$$|\rho \boldsymbol{H}_5' \hat{\boldsymbol{S}}_{kk} \boldsymbol{H}_5 - \boldsymbol{H}_5' \hat{\boldsymbol{S}}_{k0} \hat{\boldsymbol{S}}_{00}^{-1} \hat{\boldsymbol{S}}_{0k} \boldsymbol{H}_5| = 0 \,, \tag{7.14}$$

with $\hat{\boldsymbol{\rho}} = \hat{\rho}_1, \ldots, \hat{\rho}_{r_1}$. The test statistic is defined as

$$-2\ln Q(\mathcal{H}_5|\mathcal{H}_1(r)) = T\{\sum_{i=1}^{r_1} \ln(1-\hat{\rho}_i) + \sum_{i=1}^{r_2} \ln(1-\tilde{\lambda}_i) - \sum_{i=1}^{r} \ln(1-\hat{\lambda}_i)\} \,, \tag{7.15}$$

which is asymptotically distributed as χ^2 with $(N-r)r_1$ degrees of freedom.

The authors applied this test statistic to validate whether the purchasing power parity or the interest rate differential form a stationary process by themselves. This test is implemented in the function bh5lrtest() contained in the contributed package urca. In Rcode example 7.5, the results are replicated. The assumed-to-be-known partial cointegration matrices are set up as matrix objects H.51 and H.52, respectively. In the following lines, the test is applied to both of them.

Rcode 7.5 $\mathcal{H}_3$ model: Testing for partly known cointegration relations

```
H.51 <- c(1, -1, -1, 0, 0)                              1
H.52 <- c(0, 0, 0, 1, -1)                               2
H51 <- summary(bh5lrtest(z=H1, H=H.51, r=2))            3
H52 <- summary(bh5lrtest(z=H1, H=H.52, r=2))            4
```

---

	test statistic	p-value	λ_1	λ_2	λ_3	λ_4
$\mathcal{H}_{5.1}|\mathcal{H}_1(r=2)$	14.521	0.002	0.396	0.281	0.254	0.101
$\mathcal{H}_{5.2}|\mathcal{H}_1(r=2)$	1.895	0.595	0.406	0.261	0.105	0.101

Table 7.8. $\mathcal{H}_5$ model: Partly known cointegration relations

The results are exhibited in Table 7.8. The hypothesis that the PPP relation is stationary is rejected, whereas the hypothesis that the interest differential forms a stationary process cannot be rejected.

Finally, the model hypothesis $\mathcal{H}_6 : \boldsymbol{\beta} = (\boldsymbol{H}_6\boldsymbol{\varphi}, \boldsymbol{\Psi})$ has to be discussed. Recall that this hypothesis is used for testing some restrictions placed on the first r_1 cointegration relations and the remaining ones contained in $\boldsymbol{\Psi}$ are estimated freely. In contrast to the previous two model hypotheses, one cannot reduce this one to a simple eigenvalue problem. Johansen and Juselius [55] proposed a simple switching algorithm instead. The algorithm is initialized by setting $\boldsymbol{\Psi} = 0$, and the eigenvalue problem below is solved for $\boldsymbol{\varphi}$:

$$|\lambda \boldsymbol{H}_6'\hat{\boldsymbol{S}}_{kk}\boldsymbol{H}_6 - \boldsymbol{H}_6'\hat{\boldsymbol{S}}_{k0}\hat{\boldsymbol{S}}_{00}^{-1}\hat{\boldsymbol{S}}_{0k}\boldsymbol{H}_6| = 0 , \qquad (7.16)$$

which results in the eigenvalues $\hat{\lambda}_1 > \ldots > \hat{\lambda}_s > 0$ and the corresponding eigenvectors $(\hat{\boldsymbol{v}}_1, \ldots, \hat{\boldsymbol{v}}_s)$. The first partition of the cointegration relations, *i.e.*, the restricted ones, is therefore given by $\hat{\boldsymbol{\beta}}_1 = \boldsymbol{H}_6(\hat{\boldsymbol{v}}_1, \ldots, \hat{\boldsymbol{v}}_{r_1})$, although it is preliminary. The algorithm starts by fixing these values $\hat{\boldsymbol{\beta}}_1$ and by conditioning on $\hat{\boldsymbol{\beta}}_1 \boldsymbol{R}_{kt}$. It leads to the following eigenvalue problem for $\boldsymbol{\Psi}$:

$$\frac{|\boldsymbol{\Psi}'(\hat{\boldsymbol{S}}_{kk.\hat{\boldsymbol{\beta}}_1} - \hat{\boldsymbol{S}}_{k0.\hat{\boldsymbol{\beta}}_1}\hat{\boldsymbol{S}}_{00.\hat{\boldsymbol{\beta}}_1}^{-1}\hat{\boldsymbol{S}}_{0k.\hat{\boldsymbol{\beta}}_1})\boldsymbol{\Psi}|}{|\boldsymbol{\Psi}'\hat{\boldsymbol{S}}_{kk.\hat{\boldsymbol{\beta}}_1}\boldsymbol{\Psi}|} , \qquad (7.17)$$

where the product moment matrices $\hat{\boldsymbol{S}}_{ij.b}$ are given by

$$\hat{\boldsymbol{S}}_{ij.b} = \hat{\boldsymbol{S}}_{ij} - \hat{\boldsymbol{S}}_{ik}\hat{\boldsymbol{\beta}}_b(\hat{\boldsymbol{\beta}}_b'\hat{\boldsymbol{S}}_{kk}\hat{\boldsymbol{\beta}}_b)^{-1}\hat{\boldsymbol{\beta}}_b'\hat{\boldsymbol{S}}_{kj} \text{ for } i, j = 0, k \text{ and } b = 1, 2 . \qquad (7.18)$$

The solution to the eigenvalue problem 7.17 is given as Lemma 1 in Johansen and Juselius [55], and an extended exposition of eigenvalues and eigenvectors is given in appendix A.1 in Johansen [53]. Equation 7.18 yields eigenvalues $\tilde{\lambda}_1, \ldots, \tilde{\lambda}_{N-r_1}$ and eigenvectors $(\hat{\boldsymbol{u}}_1, \ldots, \hat{\boldsymbol{u}}_{N-r_1})$. Hence, the second partition of cointegration relations is given as $\hat{\boldsymbol{\beta}}_2 = (\hat{\boldsymbol{u}}_1, \ldots, \hat{\boldsymbol{u}}_{r_2})$, although it is preliminary. The second step of the algorithm consists of holding these cointegration relations fixed and conditioning on $\hat{\boldsymbol{\beta}}_2 \boldsymbol{R}_{kt}$. Hereby, a new estimate of $\hat{\boldsymbol{\beta}}_1$ is obtained by solving

$$\frac{|\boldsymbol{\varphi}'\boldsymbol{H}_6'(\hat{\boldsymbol{S}}_{kk.\hat{\boldsymbol{\beta}}_2} - \hat{\boldsymbol{S}}_{k0.\hat{\boldsymbol{\beta}}_2}\hat{\boldsymbol{S}}_{00.\hat{\boldsymbol{\beta}}_2}^{-1}\hat{\boldsymbol{S}}_{0k.\hat{\boldsymbol{\beta}}_2})\boldsymbol{H}_6\boldsymbol{\varphi}|}{|\boldsymbol{\varphi}'\boldsymbol{H}_6'\hat{\boldsymbol{S}}_{kk.\hat{\boldsymbol{\beta}}_2}\boldsymbol{H}_6\boldsymbol{\varphi}|} , \qquad (7.19)$$

which results in eigenvalues $\hat{\omega}_1, \ldots, \hat{\omega}_s$ and eigenvectors $(\hat{\boldsymbol{v}}_1, \ldots, \hat{\boldsymbol{v}}_s)$. The new estimate for $\boldsymbol{\beta}_1$ is then given by $\hat{\boldsymbol{\beta}}_1 = \boldsymbol{H}_6(\hat{\boldsymbol{v}}_1, \ldots, \hat{\boldsymbol{v}}_{r_1}$. Equations 7.17 and 7.18 are forming the switching algorithm, by calculating consecutively new sets of eigenvalues and corresponding eigenvectors until convergence is achieved; *i.e.*, the change between the values from one iteration to the next is smaller than an *a priori* given convergence criterion. Alternatively, one could iterate as long as the likelihood function:

$$L_{\max}^{-2/T} = |\hat{\boldsymbol{S}}_{00.\hat{\boldsymbol{\beta}}_1}|\prod_{i=1}^{r_2}(1 - \hat{\lambda}_i) = |\hat{\boldsymbol{S}}_{00.\hat{\boldsymbol{\beta}}_2}|\prod_{i=1}^{r_1}(1 - \hat{\omega}_i) \qquad (7.20)$$

has not achieved its maximum. Unfortunately, this algorithm does not necessarily converge to a global maximum but to a local one instead.

Last, to calculate the likelihood ratio test statistic, the eigenvalue problem:

$$|\rho\hat{\beta}_1\hat{S}_{kk}\hat{\beta}_1 - \hat{\beta}_1\hat{S}_{k0}\hat{S}_{00}^{-1}\hat{S}_{0k}\hat{\beta}_1| = 0 \qquad (7.21)$$

has to be solved for the eigenvalues $\hat{\rho}_1, \ldots, \hat{\rho}_{r_1}$. The test statistic is then given as

$$-2\ln Q(\mathcal{H}_6|\mathcal{H}_1(r)) = T\{\sum_{i=1}^{r_1}\ln(1-\hat{\rho}_i) + \sum_{i=1}^{r_2}\ln(1-\tilde{\lambda}_i) - \sum_{i=1}^{r}\ln(1-\hat{\lambda}_i)\}, \quad (7.22)$$

which is asymptotically distributed as χ^2 with $(N - s - r_2)r_1$ degrees of freedom.

This test statistic is implemented as function bh6lrtest() in the contributed package urca. Beside the $\mathcal{H}_1$ object and the restriction matrix, the total number of cointegration relations, the number of restricted relationships, as well as the convergence value and the maximum number of iterations enter as functional arguments. The convergence criterium is defined as the vector norm of $\tilde{\lambda}$.

Because the test result of the model hypothesis $\mathcal{H}_5$ indicated that the purchasing power parity does not hold in the strict sense, the authors applied this test to see whether a more general linear but still stationary combination of p_1, p_2, and e_{12} exists. That is, the question now is whether a more general cointegration vector of the form $(a, b, c, 0, 0)$ does yield a stationary process. This restriction can be casted into the following matrix $\boldsymbol{H}_6$:

$$\boldsymbol{H}_6 = \begin{bmatrix} 1 & 0 & 0 \\ 0 & 1 & 0 \\ 0 & 0 & 1 \\ 0 & 0 & 0 \\ 0 & 0 & 0 \end{bmatrix}.$$

The application of this test is provided in Rcode example 7.6 and its results are depicted in Table 7.9. The test statistic is not significant at the 1% level. Please note that compared with the results in Johansen and Juselius [55], the algorithm converged to slightly other values for the second, third, and fourth eigenvalue.

Rcode 7.6 $\mathcal{H}_6$ model: Testing of restrictions on r_1 cointegration relations

---

```
H.6 <- matrix(rbind(diag(3), c(0, 0, 0), c(0, 0, 0)), nrow      1
   =5, ncol=3)
H6 <- summary(bh6lrtest(z=H1, H=H.6, r=2, r1=1))               2
```

---

	test statistic	p-value	$\hat{\lambda}_1$	$\hat{\lambda}_2$	$\hat{\lambda}_3$	$\hat{\lambda}_4$
$\mathcal{H}_6 | \mathcal{H}_1 (r=2)$	4.931	0.026	0.407	0.281	0.149	0.091

Table 7.9. $\mathcal{H}_6$ model: Restrictions on r_1 cointegration relations

7.2 VECMs and Structural Shift

In Section 5.1, the implications for the statistical inference of unit root tests in light of structural breaks have been discussed. The pitfalls of falsely concluding nonstationarity in the data can also be encountered in the case of VECMs. The flipside would be a wrongly accepted cointegration relation, whereas some or all underlying series behave like an AR(1)–process with a structurual break. Lütkepohl *et al.* [61] proposed a procedure for estimating a VECM, in which the structural shift is a simple shift in the level of the process. Hereby, the break date is estimated first. Next, the deterministic part including the size of the shift is estimated and the data are adjusted accordingly. Finally, a Johansen–type test for determining the cointegration rank can be applied to these adjusted series.

They assume that the $(N \times 1)$ vector process $\{y_t\}$ is generated by a constant, a linear trend, and level shift terms:

$$y_t = \mu_0 + \mu_1 t + \delta d_{t\tau} + x_t \,, \qquad (7.23)$$

where $d_{t\tau}$ is a dummy variable defined by $d_{t\tau} = 0$ for $t < \tau$ and $d_{t\tau} = 1$ for $t \geq \tau$. The shift assumed that the shift point τ is unknown and is expressed as a fixed fraction of the sample size:

$$\tau = [T\lambda] \text{ with } 0 < \underline{\lambda} \leq \lambda \leq \overline{\lambda} < 1 \,, \qquad (7.24)$$

where $\underline{\lambda}$ and $\overline{\lambda}$ define real numbers and the $[\cdot]$ defines the integer part. The meaning of Equation 7.24 is that the shift might neither occur at the very beginning nor at the very end of the sample.
Furthermore, it is assumed that the process $\{x_t\}$ can be represented as a VAR(p) and that the components are at most $I(1)$ and cointegrated with rank r.

The estimation of the break point is based on the following regressions:

$$y_t = \nu_0 + \nu_1 t + \delta d_{t\tau} + A_1 y_{t-1} + \ldots + A_p y_{t-p} + \varepsilon_{t\tau} \text{ for } t = p+1, \ldots, T \,, \quad (7.25)$$

where A_i with $i = 1, \ldots, p$ assign the $(N \times N)$ coefficient matrices and ε_t is the spherical N-dimensional error process. It should be noted that other exogenous regressors, like seasonal dummy variables, can also be included in Equation 7.25.

The estimator for the break point $\hat{\tau}$ is then defined as

$$\hat{\tau} = \arg\min_{\tau \in \mathfrak{T}} \det\left(\sum_{t=p+1}^{T} \hat{\varepsilon}_{t\tau} \hat{\varepsilon}'_{t\tau} \right), \tag{7.26}$$

where $\mathfrak{T} = [T\underline{\lambda}, T\overline{\lambda}]$ and $\hat{\varepsilon}_{t\tau}$ are the least-squares residuals of Equation 7.25. The integer count of the interval $\mathfrak{T} = [T\underline{\lambda}, T\overline{\lambda}]$ determines how many regressions have to be run with the correponding step dummy variables $d_{t\tau}$ and for how many times the determinant of the product moment matrices of $\hat{\varepsilon}_{t\tau}$ have to be calculated. The minimal one is the one that selects the most likely break point.

Once the break point $\hat{\tau}$ is estimated, the data are adjusted according to

$$\hat{x}_t = y_t - \hat{\mu}_0 - \hat{\mu}_1 t - \hat{\delta} d_{t\hat{\tau}}. \tag{7.27}$$

This method is included as function `cajolst()` in the contributed package urca. By applying this function, an object of class `ca.jo` is generated. The adjusted series are in the slot x, and the estimate of the break point is stored in the slot bp. Instead of using the test statistic as proposed in Lütkepohl *et al.* [61] with critical values provided in Lütkepohl and Saikkonen [60] the test statistic:

$$LR(r) = T \sum_{j=r+1}^{N} \ln(1 + \hat{\lambda}_j) \tag{7.28}$$

has been implemented with critical values from Trenkler [97] in the function `cajolst()`. The advantage is that in the latter source, the critical values are provided more extensively and precisely.

In Rcode example 7.7, this method has been applied to estimate a money demand function for Denmark as in Johansen and Juselius [54]. For a better comparison, the results for the nonadjusted data are also given in Table 7.10.

Rcode 7.7 $\mathcal{H}_1$ model: Inference on cointegration rank for Danish money demand function allowing for structural shift

---

```
data(denmark)                                                        1
sjd <- denmark[, c("LRM", "LRY", "IBO", "IDE")]                      2
sjd.vecm <- summary(ca.jo(sjd, constant = TRUE, type = "           3
    eigen", K = 2, spec = "longrun", season = 4, ctable = "
    A3"))
lue.vecm <- summary(cajolst(sjd, season=4))                          4
```

	test statistic	10%	5%	1%
r <= 3	2.35	7.56	9.09	12.74
r <= 2	6.34	13.78	15.75	19.83
r <= 1	10.36	19.80	21.89	26.41
r = 0	30.09	25.61	28.17	33.12

Table 7.10. Money demand function for Denmark: Maximal eigenvalue statistic, nonadjusted data

	test statistic	10%	5%	1%
r <= 3	3.15	5.42	6.79	10.04
r <= 2	11.62	13.78	15.83	19.85
r <= 1	24.33	25.93	28.45	33.76
r = 0	42.95	42.08	45.20	51.60

Table 7.11. Money demand function for Denmark: Trace statistic, allowing for structural shift

For the nonadjusted data, the hypothesis of one cointegration relation cannot be rejected for a significance level of 5%. If one allows for a structural shift in the data, however, one cannot reject the hypothesis of no cointegration as indicated by the results in Table 7.11. The shift occurred most likely in 1975:Q4. Therefore, a VAR in differences with an intervention dummy for that period might be a more suitable model to describe the data-generated process.

Summary

In the last chapter of the book, likelihood-based inference in cointegrated vector autoregressive models has been presented. It has been shown how to determine the cointegration rank and, dependent on that outcome, how to specify and test the validity of restrictions placed on the cointegrating and the weighting matrix. This methodology offers the researcher a powerful tool to investigate the relationships in a system of cointegrated variables more thoroughly compared with the single equation methods presented in Chapter 6. Furthermore, it has been shown how one can employ this methodology in light of a structural shift at an unknown point in time.

Exercises

1. Consider the data sets `finland` and `denmark` in the contributed package `urca`. Specify for each country a VECM that mirrors a real money demand function.
2. Reconsider the data sets `Raotbl1` and `Raotbl1` in the contributed package `urca`. Now specify for each monetary aggregate a VECM and compare your findings with the results from Exercise 1 in Chapter 6.
3. Reconsider the data set `Raotbl6` in the contributed package `urca`. Now, specify a VECM Phillips-Curve model as in Mehra [68]. Discuss your findings compared with your results of Exercise 2 in Chapter 6.

8

Appendix

8.1 Used CRAN Packages

Name	Title	Version	Date/Citation
chron	Chronological objects which can handle dates and times	2.2-34	2005-02-26 [50]
dse1	Dynamic Systems Estimation (time series package)	2005.1-1	2005.1-1 [73]
fBasics	Financial Software Collection - fBasics	201.10059	1996-2005 [101]
fracdiff	Fractionally differenced ARIMA (p,d,q) models	1.1-1	2004-01-12 [29]
fSeries	Financial Software Collection - fSeries	201.10059	1996-2005 [102]
lmtest	Testing Linear Regression Models	0.9-10	2005-01-21 [105]
Rcmdr	R Commander	1.0-0	2005/4/19 [28]
strucchange	Testing for Structural Change	1.2-9	2005-04-06 [106]
tseries	Time series analysis and computational finance	0.9-26	2005-04-20 [96]
urca	Unit root and cointegration tests for time series data	0.8-2	2005-04-24 [79]
uroot	Unit root tests and graphics for seasonal time series.	1.2	2005/3/3 [59]
xtable	Export tables to LaTeX or HTML	1.2-5	2004/12/01 [12]
zoo	Z's ordered observations	0.9-1	2004-12-17 [104]

Table 8.1. Overview of used packages

8.2 Time Series Data

Several possibilities deal with time series data in R. First, the class *ts* in the base distribution is well suited for handling regular spaced time series data. In Rcode example 8.1, it is shown how to assign the range as well as the frequency to the data frame `finland` contained in the package `urca`. Objects of class *ts* own a time series property that can be shown by the function `tsp()`. The time component of an object of class *ts* can be retrieved with the function `time()`. Finally, subsetting a time series object to a narrower sample range is accomplished by using the `window()` function.

Rcode 8.1 Time series objects of class "ts"

```
# time series handling in R                                       1
library(urca)                                                     2
data(finland)                                                     3
str(finland)                                                      4
# utilisation of time series class 'ts' in base package          5
fin.ts <- ts(finland, start=c(1958, 2), end=c(1984, 3),          6
    frequency=4)
str(fin.ts)                                                       7
# time series properties of fin.ts                               8
tsp(fin.ts)                                                       9
time(fin.ts)[1:10]                                               10
# Creating a sub sample                                          11
finsub.ts <- window(fin.ts, start=c(1960, 2), end=c(1979, 3)    12
    )
tsp(finsub.ts)                                                   13
```

Second, mostly encountered in financial econometric applications is the case of irregularily spaced series with respect to time. Four contributed packages do exist in R that particularly address this issue, *i.e*, fBasics, its, tseries, and zoo. Although these packages do differ on how certain functionalities and classes are defined, building unions, intersections, and the merging of objects can be achieved with all of them, although the package its is the most mature. The functions `its()` and `timeSeries()` in the packages its and fBasics have been implemented as S4 classes, whereas the functions `irts()` and `zoo()` in the packages tseries and zoo are S3 classes for irregularily spaced observations. The advantage of zoo compared with the other functionalities is that time information can be of almost any class, whereas in the other implementations, it needs to be of class POSIXct. The handling of irregularily time series in the package fBasics resembles those in the finmetrics package of S-PLUS.

8.3 Technicalities

This book has been typeset in LaTeX. Text editor `Emacs/ESS` has been used. The indices have been generated with the program `makeindex` and the bibliography with `BiBTeX`. The flow chart (see Figure 2.3) has been produced with the program `flow`. The following LaTeX packages have been used: `amsmath`, `amssymb`, `bm`, `float`, `graphicx`, `index`, `listings`, `multicol`, `paralist`, and `sweave`.

All `RCode` examples have been processed as `Sweave`-files. Therefore, the proper working of the `R` commands is guaranteed. Where possible the results are exhibited as tables by making use of the function `xtable()` contained in the contributed package of the same name. The examples have been processed under R version 2.1.1 on an i686 PC with Linux as the operating system and kernel 2.6.5-7.151-default. Linux is a registered trademark of Linus Torvalds (Helsinki, Finland), the original author of the Linux kernel. All contributed packages have been updated before publishing and are listed in Table 8.1 in the Appendix.

9

Abbreviations, Nomenclature, and Symbols

Abbreviations:

ACF	autocorrelation function
ADF	augmented Dickey–Fuller
ADL	autoregressive-distributed lag
AIC	Akaike information criteria
AR	autoregression
ARIMA	autoregressive integrated moving average
ARFIMA	autoregressive fractionally integrated moving average
ARMA	autoregressive-moving average
BIC	Bayesian information criteria
CI(d, b)	cointegrated of order d, b
CRDW	cointegrating regression Durbin–Watson statistic
DF	Dickey–Fuller
DGP	data-generating process
ECM	error-correction model / mechanism
ERS	Elliott, Rothenberg and Stock
GNP	Gross National Product
HEGY	Hylleberg, Engle, Granger and Yoo
I(d)	integrated of order d
i.d.	independently distributed
i.i.d.	independently and identically distributed
LB	Ljung–Box Portmanteau
LM	Lagrange multiplier
KPSS	Kwiatkowski, Phillips, Schmidt, and Shin
MA	moving average
NI	near integrated
OLS	ordinary least-squares

PACF	partial autocorrelation function
PP	Phillips and Perron
SC	Schwarz criteria
SI	seasonally integrated
SP	Schmidt and Phillips
T	sample size or last observation in a time series
VAR	vector autoregression
var	variance
VECM	vector error correction model
ZA	Zivot and Andrews

Nomenclature:

Bold lower case: $\boldsymbol{y}, \boldsymbol{\alpha}$	vectors
Bold upper case: $\boldsymbol{Y}, \boldsymbol{\Gamma}$	matrices
Greek letters: α, β, γ	population values (parameters)
Greek letters with ˆ or ˜	sample values (estimates or estimators)
Y, y	endogenous variables
X, x, Z, z	exogenous or pre-determined variables
L	Lag operator, defined as $Lx_t = x_{t-1}$
Δ	first-difference operator: $\Delta x_t = x_t - x_{t-1}$

Symbols:

$\perp$	Orthogonality sign
$\cap$	intersection
$\in$	set membership
$\dim()$	Dimension
$\Gamma()$	Gamma function
i	complex number
$\mathcal{H}$	Hypotheses
$\mathcal{N}$	Normal distribution
$rk()$	Rank of a matrix
$sp()$	Space
$tr()$	Trace of a matrix

List of Tables

List of Figures

List of Rcode

References

1. H. Akaike. Likelihood of a model and information criteria. *Journal of Econometrics*, 16:3–14, 1981.
2. A. A. Anis and E. H. Lloyd. The expected value of the adjusted rescaled Hurst range of independent normal summands. *Biometrika*, 63:111–116, 1976.
3. K. Aydogan and G. G. Booth. Are there long cycles in common stock returns? *Southern Economic Journal*, 55:141–149, 1988.
4. R. T. Baillie. Long memory processes and fractional integration in econometrics. *Journal of Econometrics*, 73:5–59, 1996.
5. A. K. Bera and C. M. Jarque. Efficient tests for normality, heteroscedasticity, and serial independence of regression residuals. *Economic Letters*, 6:255–259, 1980.
6. A. K. Bera and C. M. Jarque. An Efficient Large-Sample Test for Normality of Observations and Regression Residuals. Working Papers in Econometrics 40, Australian National University, Canberra, 1981.
7. S. Beveridge and C. R. Nelson. A new approach to decomposition of economic time series into permanent and transitory components with particular attention to measurement of the "business cycle". *Journal of Monetary Economics*, 7:151–174, 1981.
8. G. E. P. Box and G. M. Jenkins. *Time Series Analysis: Forecasting and Control.* Holden-Day, San Francisco, revised edition, 1976.
9. J. Y. Campbell and P. Perron. Pitfalls and Opportunities: What Macroeconomists should know about Unit Roots. In *NBER Macroeconomic Annual 1991*, pages 141–218. National Bureau of Economic Research, Cambridge, M.A., 1991.
10. F. Canova and B. E. Hansen. Are seasonal patterns constant over time? A test for seasonal stationarity. *Journal of Business and Economic Statistics*, 13:237–252, 1995.
11. J. M. Chambers. *Programming with Data: A Guide to the S Language.* Springer-Verlag, New York, 3rd edition, 2000.
12. D. B. Dahl and with contributions from many others. *xtable: Export Tables to LaTeX or HTML*, 2004. R package version 1.2-4.
13. P. Dalgaard. *Introductory Statistics with R.* Springer-Verlag, New York, 2002.
14. R. Davidson and J. G. MacKinnon. *Estimation and Inference in Econometrics.* Oxford University Press, Oxford, U.K., 1992.

15. R. B. Davies and D. S. Harte. Tests for Hurst effects. *Biometrika*, 74:95–102, 1987.

16. D. A. Dickey and W. A. Fuller. Distributions of the estimators for autoregressive time series with a unit root. *Journal of the American Statistical Association*, 74:427–431, 1979.

17. D. A. Dickey and W. A. Fuller. Likelihood ratio statistics for autoregressive time series with a unit root. *Econometrica*, 49:1057–1072, 1981.

18. D. A. Dickey, D. P. Hasza, and W. A. Fuller. Testing for unit roots in seasonal time series. *Journal of the American Statistical Association*, 5:355–367, 1984.

19. D. A. Dickey and S. G. Pantula. Determining the order of differencing in autoregressive process. *Journal of Business & Economic Statistics*, 5(4):455–461, 1987.

20. F. X. Diebold and G. D. Rudebusch. Long memory and persistence in aggregate output. *Journal of Monetary Economics*, 24:189–209, 1989.

21. J. Durbin and G. S. Watson. Testing for serial correlation in least-squares regression I. *Biometrika*, 37:409–428, 1950.

22. J. Durbin and G. S. Watson. Testing for serial correlation in least-squares regression II. *Biometrika*, 38:159–178, 1951.

23. J. Durbin and G. S. Watson. Testing for serial correlation in least-squares regression III. *Biometrika*, 58:1–19, 1971.

24. G. Elliott, T. J. Rothenberg, and J. H. Stock. Efficient tests for an autoregressive unit root. *Econometrica*, 64(4):813–836, Jul 1996.

25. R. Engle and S. Yoo. Forecasting and testing in cointegrated systems. *Journal of Econometrics*, 35:143–159, 1987.

26. R. F. Engle and C. W. J. Granger. Co-integration and error correction: Representation, estimation, and testing. *Econometrica*, 55(2):251–276, March 1987.

27. R. F. Engle, C. W. J. Granger, and J. Hallman. Merging short- and long-run forecasts: An application of seasonal co-integration to monthly electricity sales forecasting. *Journal of Econometrics*, 40:45–62, 1988.

28. J. Fox. *Rcmdr: R Commander*, 2004. R package version 0.9-14.

29. C. Fraley, F. Leisch, and M. Maechler. *fracdiff: Fractionally Differenced ARIMA (p,d,q) Models*, 2004. R package version 1.1-1, S original by Fraley, C., R Port by Leisch, F. and since 2003-12 maintainer Mächler, M.

30. P. H. Franses and B. Hobijn. Critical values for unit root tests in seasonal time series. *Journal of Applied Statistics*, 24:25–48, 1997.

31. R. Frisch and F. V. Waugh. Partial time regressions as compared with individual trends. *Econometrica*, 1:387–401, 1933.

32. W. A. Fuller. *Introduction to Statistical Time Series*. John Wiley & Sons Inc., New York, 1976.

33. J. Geweke and S. Porter-Hudak. The estimation and application of long memory time series. *Journal of Time Series Analysis*, (4):221–238, 1983.

34. C. W. J. Granger. Long memory relationships and the aggregation of dynamic models. *Journal of Econometrics*, 14:227–238, 1980.

35. C. W. J. Granger. Some properties of time series data and their use in econometric model specification. *Journal of Econometrics*, 16:150–161, 1981.

36. C. W. J. Granger and R. Joyeux. An introduction to long-memory time series models and fractional differencing. *Journal of Time Series Analysis*, 1:15–29, 1980.

37. C. W. J. Granger and P. Newbold. Spurious regressions in econometrics. *Journal of Econometrics*, 2:111–120, 1974.

38. J. D. Hamilton. *Time Series Analysis*. Princeton University Press, Princeton, N.J., 1994.

39. J. Haslett and A. E. Raftery. Space-time modelling with long-memory dependence: Assessing Ireland's wind power resource (with Discussion). *Applied Statistics*, 38:1–50, 1989.

40. D. P. Hasza and W. A. Fuller. Testing for nonstationary parameter specifications in seasonal time series models. *The Annals of Statistics*, 10:1209–1216, 1982.

41. D. F. Hendry. Econometrics: Alchemy or science? *Economica*, 47:387–406, 1980.

42. D. F. Hendry. Econometric modelling with cointegrated variables: An overview. *Oxford Bulletin of Economics and Statistics*, 48(3):201–212, 1986.

43. D. F. Hendry. The Nobel Memorial Prize for Clive W. J. Granger. *Scandinavian Journal of Economics*, 106(2):187–213, 2004.

44. D. F. Hendry and G. J. Anderson. Testing dynamic specification in small simultaneous systems: An application to a model of building society behaviour in the United Kingdom. In M. D. Intriligator, editor, *Frontiers in Quantitative Economics*, volume 3, pages 361–383. North-Holland, Amsterdam, 1977.

45. D. Holden and R. Perman. *Unit Roots and Cointegration for the Economist*, chapter 3. In Rao [84], 1994.

46. R. H. Hooker. Correlation of the marriage-rate with trade. *Journal of the Royal Statistical Society*, 64:485–492, 1901.

47. J. R. M. Hosking. Fractional differencing. *Biometrika*, 68(1):165–176, 1981.

48. H. Hurst. Long term storage capacity of reservoirs. *Transactions of the American Society of Civil Engineers*, 116:770–799, 1951.

49. S. Hylleberg, R. F. Engle, C. W. J. Granger, and B. S. Yoo. Seasonal integration and cointegration. *Journal of Econometrics*, 69:215–238, 1990.

50. James, D. (S original) and Hornik, K. (R port). *chron: Chronological objects which can handle dates and times*, 2004. R package version 2.2-33.

51. S. Johansen. Statistical analysis of cointegration vectors. *Journal of Economic Dynamics and Control*, 12:231–254, 1988.

52. S. Johansen. Estimation and hypothesis testing of cointegration vectors in Gaussian vector autoregressive models. *Econometrica*, 59(6):1551–1580, November 1991.

53. S. Johansen. *Likelihood-Based Inference in Cointegrated Vector Autoregressive Models*. Advanced Texts in Econometrics. Oxford University Press, Oxford, 1995.

54. S. Johansen and K. Juselius. Maximum likelihood estimation and inference on cointegration - with applications to the demand for money. *Oxford Bulletin of Economics and Statistics*, 52(2):169–210, 1990.

55. S. Johansen and K. Juselius. Testing structural hypothesis in a multivariate cointegration analysis of the PPP and the UIP for UK. *Journal of Econometrics*, 53:211–244, 1992.

56. D. Kwiatkowski, P. C. B. Phillips, P. Schmidt, and Y. Shin. Testing the null hypothesis of stationarity against the alternative of a unit root: How sure are we that economic time series have a unit root? *Journal of Econometrics*, 54:159–178, 1992.

57. G. M. Ljung and G. E. P. Box. On a measure of lack of fit in time series models. *Biometrika*, 65:553–564, 1978.

58. A. W. Lo. Long-term memory in stock market prices. *Econometrica*, 59(5):1279–1313, September 1991.

59. Javier López-de Lacalle and Ignacio Díaz-Emparanza. *uroot: Unit root tests and graphics for seasonal time series.*, 2004. R package version 1.0.

60. H. Lütkepohl and P. Saikkonen. Testing for the cointegration rank of a VAR process with a time trend. *Journal of Econometrics*, 95:177–198, 2000.

61. H. Lütkepohl, P. Saikkonen, and C. Trenkler. Testing for the cointegrating rank of a VAR with level shift at unknown time. *Econometrica*, 72(2):647–662, March 2004.

62. J. MacKinnon. *Critical Values for Cointegration Tests*, chapter 13. Advanced Texts in Econometrics. Oxford, UK: Oxford University Press, 1991.

63. B. B. Mandelbrot. Statistical methodology for non periodic cycles: From the covariance to R/S analysis. *Annals of Economic and Social Measurement*, 1:259–290, 1972.

64. B. B. Mandelbrot. A fast fractional gaussian noise generator. *Water Resources Research*, 7:543–553, 1975.

65. B. B. Mandelbrot and J. Wallis. Noah, Joseph and operational hydrology. *Water Resources Research*, 4:909–918, 1968.

66. B. B. Mandelbrot and J. Wallis. Robustness of the rescaled range R/S in the measurement of noncyclical long-run statistical dependence. *Water Resources Research*, 5:967–988, 1969.

67. A. I. McLeod and K. W. Hipel. Preservation of the rescaled adjusted range, 1: A reassessment of the Hurst phenomenon. *Water Resources Research*, 14:491–508, 1978.

68. Y. P. Mehra. *Wage Growth and the Inflation Process: An Empirical Approach*, chapter 5. In Rao [84], 1994.

69. C. R. Nelson and C. I. Plosser. Trends and random walks in macroeconomic time series. *Journal of Monetary Economics*, 10:139–162, 1982.

70. W. Newey and K. West. A simple positive definite, heteroscedasticity and autocorrelation consistent covariance matrix. *Econometrica*, 55:703–705, 1987.

71. D. R. Osborn, A. P. L. Chui, J. P. Smith, and C. R. Birchenhall. Seasonality and the order of integration for consumption. *Oxford Bulletin of Economics and Statistics*, 54:361–377, 1988.

72. S. Ouliaris, J. Y. Park, and P. C. B. Phillips. Testing for a unit root in the presence of a maintained trend. In B. Raj, editor, *Advances in Econometrics and Modelling*, pages 7–28. Kluwer Academic Publishers, Dordrecht, 1989.

73. G. Paul. *dse1: Dynamic Systems Estimation (time series package)*, 2004. R package version 2004.10-1.

74. P. Perron. Trends and random walks in macroeconomic time series. *Journal of Economic Dynamics and Control*, 12:297–332, 1988.

75. P. Perron. The Great Crash, the Oil Price Shock, and the Unit Root Hypothesis. *Econometrica*, 57(6):1361–1401, November 1989.

76. P. Perron. Testing for a unit root in a time series with a changing mean. *Journal of Business & Economic Statistics*, 8(2):153–162, April 1990.

77. P. Perron. Erratum: The Great Crash, the Oil Price Shock and the Unit Root Hypothesis. *Econometrica*, 61(1):248–249, January 1993.

78. P. Perron and T. J. Vogelsang. Testing for a unit root in a time series with a changing mean: corrections and extensions. *Journal of Business & Economic Statistics*, 10:467–470, 1992.

79. B. Pfaff. *urca: Unit root and cointegration tests for time series data*, 2004. R package version 0.6-1.

80. P. C. B. Phillips. Understanding Spurious Regressions in Econometrics. Cowles Foundation for Research in Economics, Yale University, Cowles Foundation Paper 667, 1986.

81. P. C. B. Phillips and S. Ouliaris. Asymptotic Properties of residual based tests for cointegration. *Econometrica*, 58:165–193, 1990.

82. P. C. B. Phillips and P. Perron. Testing for a unit root in time series regression. *Biometrika*, 75(2):335–346, 1988.

83. R Development Core Team. *R: A Language and Environment for Statistical Computing*. R Foundation for Statistical Computing, Vienna, Austria, 2004. ISBN 3-900051-07-0.

84. B. B. Rao, editor. *Cointegration for the Applied Economist*. London, UK: The MacMillan Press Ltd., 1994.

85. P. M. Robinson. Efficient tests of nonstationary hypotheses. *Journal of the American Statistical Association*, 89(428):1420–1437, December 1994.

86. J. D. Sargan. Wages and prices in the United Kingdom: A study in econometric methodology (with Discussion). In P. E. Hart, G. Mills, and J. K. Whitaker, editors, *Econometric Analysis for National Economic Planning*, volume 16 of *Colsion Papers*, pages 25–63. Butterworth Co., London, 1964.

87. J. D. Sargan and A. Bhargava. Testing residuals from least squares regression for being generated by the Gaussian random walk. *Econometrica*, 51:153–174, 1983.

88. P. Schmidt and P. C. B. Phillips. LM tests for a unit root in the presence of deterministic trends. *Oxford Bulletin of Economics and Statistics*, 54(3):257–287, 1992.

89. P.C. Schotman and H. K. van Dijk. On Bayesian routes to unit roots. *Journal of Applied Econometrics*, 6:387–401, 1991.

90. H. Schwarz. Estimating the dimension of a model. *The Annals of Statistics*, 6:461–464, 1978.

91. S. S. Shapiro and M. B. Wilk. An analysis of variance test for normality (complete samples). *Biometrika*, 52:591–611, 1965.

92. S. S. Shapiro, M. B. Wilk, and H. J. Chen. A comparative study of various tests of normality. *Journal of the American Statistical Association*, 63:1343–1372, 1968.

93. F. B. Sowell. Modeling long run behaviour with fractional ARIMA model. *Journal of Monetary Economics*, 29:277–302, 1992.

94. A. Spanos. *Statistical Foundations of Econometric Modelling*. Cambridge University Press, 1986.

95. J. H. Stock. Asymptotic properties of least squares estimators of cointegrating vectors. *Econometrica*, 55:1035–1056, 1987.

96. A. Trapletti and K. Hornik. *tseries: Time Series Analysis and Computational Finance*, 2004. R package version 0.9-24.

97. C. Trenkler. A new set of critical values for systems cointegration tests with a prior adjustment for deterministic terms. *Economics Bulletin*, 3(11):1–9, 2003.

98. W. N. Venables and B. D. Ripley. *Modern Applied Statistics with S*. Springer-Verlag Inc., New York, 4th edition, 2002.

99. K. F. Wallis. Seasonal adjustment and relations between variables. *Journal of the American Statistical Association*, 69:18–31, 1974.

100. H. White. *Asymptotic Theory for Econometricians*. Academic Press, New York, 1984.
101. D. Würtz and et al. *fBasics: Financial Software Collection-fBasics*, 2004. R package version 200.10058.
102. D. Würtz and et al. *fSeries: Financial Software Collection-fSeries*, 2004. R package version 200.10058.
103. G. U. Yule. Why do we sometimes get nonsense-correlations between time series? A study in sampling and the nature of time series. *Journal of the Royal Statistical Society*, 89:1–64, 1926.
104. A. Zeileis and G. Grothendieck. *zoo: Z's Ordered Observations*, 2004. R package version 0.2-0.
105. A. Zeileis and T. Hothorn. Diagnostic checking in regression relationships. *R News*, 2(3):7–10, 2002.
106. A. Zeileis, F. Leisch, K. Hornik, and C. Kleiber. strucchange: An R package for testing for structural change in linear regression models. *Journal of Statistical Software*, 7(2):1–38, 2002.
107. E. Zivot and D. W. K. Andrews. Further evidence on the Great Crash, the Oil-Price Shock, and the Unit-Root Hypothesis. *Journal of Business & Economic Statistics*, 10(3):251–270, July 1992.

Name Index

Function Index

Index